THERE ARE NO ISLANDS

THERE ARE NO ISLANDS

The Concerns and Potentials of Continuing Education

By

ROBERT E. SHARER

THE CHRISTOPHER PUBLISHING HOUSE
NORTH QUINCY, MASSACHUSETTS
02171

Library of Congress Catalog Card Number 69-16276
SBN 8158-0025-8

PRINTED IN
THE UNITED STATES OF AMERICA

Dedicated
to
My Wife Gwendolyn
whose encouraging suggestions were
a continuing support and valuable assistance

FOREWORD

There Are No Islands is not the result of a momentary impulse. The convictions about the present importance and the yet to be realized potentialities of continuing education, which comprise the fabric of this book, evolved during 47 years of participation in American education by the author. He has devoted the past 26 years, 1943-69, to continuing education.

In the United States educating adults is characterized by unusual diversity in form, content, quality, techniques and goals. This has proved to be both weakness and strength. It has resulted in confusion in nomenclature. No single term is generally used. Adult education enjoys a slight advantage because of its international acceptance. Among other terms currently used are: adjunct education (added but separate; attached but subordinate); continuing education (prolongation); extra-mural education (outside of, beyond); para-education (beyond, alongside of); and life-long learning (coincident with the life-span).

Continuing education is used in *There Are No Islands* in all references to the education of adults. Historically it has been used in the United States for almost a half century. In 1926, the governing board of Michigan State University (at that time called Michigan Agricultural College) formally established the Department of Continuing Education. Scores of institutions have since adopted continuing education in preference to other terms.

The basic reason for preferring continuing education is expressed in the beginning paragraphs of Chapter II.

> The term continuing education possesses a fascinating intrinsic quality. Continuing education implies "learning in progress" at the present time, and learning going on and on as long as the mind of the individual remains capable of comprehension and application.
>
> As long as a man lives he remains an uncompleted person. "We will BE what we WILL to be" is the prime motivation for continuing education. No person even approaches spiritual and intellectual completion except by a lifetime of learning.

The author believes continuing education is very important. Its purposes reflect the interests and needs of individuals, and those of society. He contends that the potential significance and power of continuing education have yet to be realized. He pleads in this book for leadership of unusual quality and effectiveness by men and women who influence and direct continuing education.

Ideas cannot be defeated by military or political powers. They can be either subverted or developed by use of education. The management of continuing education within a nation has frightening possibilities for enslavement of men's minds. It has unrealized potentialities for freeing humanity from the shackles of history—prejudice, injustice, ignorance and fear.

No individual or group of individuals may exempt themselves from responsibility for determining the future quality of continuing education in the United States, and for whom shall eventually "manage" it and for what purposes. It is the concern of young adults, of mature men and women, and of older individuals living as retirees.

The author admits that education is not a panacea to cure the ills of society. He contends that it is one of several effective tools to encourage the building of a better society.

There is a hope expressed in *There Are No Islands* that men and women engaged in American continuing education will devote less effort and resources to the trivial and peripheral activities within contemporary continuing education and exert more effective leadership to those programs which are concerned with the quality of living in society—Today and Tomorrow.

TABLE OF CONTENTS

THERE ARE NO ISLANDS

Chapter I

THERE ARE NO ISLANDS

Continuing Education in a Turbulent World

Edna St. Vincent Millay's phrase, "There are no islands any more," is no mere poetic lament for those earlier times when men might find convenient and frequent sanctuary for mind and soul from the press of other humans and from the stresses of living. It is a disquietingly precise definition of the world in which you and I are living.

Today the world is crowded—crowded with people of all ages, races, colors, beliefs; packed with proliferating agencies, institutions, organizations; filled up with things, gadgets, inventions; crammed with events. There are no oceans to protect nations from each other; no frontiers for the dissatisfied to escape into; no buffers of empty space between neighbors and communities; no "islands of refuge" from the exacerbations of noise, traffic, smog, tension and fear of extinction.

Our world, which has been moving sedately through the slow centuries, no longer requires a million years to invent fire and speech and a thousand years to accept a new idea. Our world is "dizzy," suffering with change-induced vertigo. The very word "change" is inadequate so we employ terms like "explosion" and "volcanic transformation" to describe what is happening in science and technology; to our population, transportation, government and basic beliefs.

We live in a schizophrenic world, hauled and pulled by conflicting ideas and blatant propaganda, bedeviled and perplexed by problems and issues of a magnitude and consequence which never before confronted mankind. Faced with possible nuclear war we consciously and subconsciously worry about survival; physical biological survival, survival of our ideas and ideals—of our way of life. Our religions are being subjected to the impact of space exploration and we have barely crossed the threshold of eventual conquest of the universe beyond. Rampaging militant communism is forcing us to defend our basic beliefs about democracy. Daily we face problems caused by urbanization, industrial automation, resurgent nationalism and similar social, economic and technological upheavals. Perhaps as a kind of defense we develop a social myopia, or an apathy toward events. We become overly concerned, worried, fearful and frustrated.

Change is inevitable, unremitting, perpetual. In some ages of society it is slow, in others accelerated, but always we have change. Individuals may react to change in one of five ways. They may resist, adapt to, accept, control or cause change. The concern of continuing education is not with change per se, but how to enable adults to choose intelligently which of these five courses of action they should follow, and then to provide them with facts and skills which make their chosen course of action effective and productive.

Primitive man knew little about himself or about the earth which was his home, practically nothing about the universe. He lived in fear and terror of the unknown, in discomfort and cold with too little food and very little light to push back the bleak darkness of long nights. His days were hard, his years were few. Less than 3,000 years ago men living on a mountain peninsula jutting into the Aegean Sea began to ask questions. Pericles persuaded the famous Persian scientist Anaxagoras to come to Athens to teach the Athenians. One of his ideas startled his adult pupils. He claimed that the sun was far larger than they believed, that it was in fact as large as all Greece. Instead of believing him his contemporaries derided him.

Today man no longer lives in terror of the phenomena of the universe. We have learned that the sun is more than a thousand times larger than our earth and millions of miles distant in space. We know our solar system is but a speck of dust in an infinitely huge expanding universe of billions of galaxies and systems of galaxies. We know that matter is composed of ultramicroscopic tiny bits of matter we call atoms. We have learned that each atom is a complex electronic system in miniature; composed of 30 particles, some have a lifetime of only millionths of a second, others are eternal. These facts merely spur scientists to investigate further and encourage individuals to seek more knowledge.

Before 1800 it was entirely possible for any normal person to master the sum total of human knowledge within the years of an average lifetime. During the centuries before 1800 the sum total of human knowledge doubled about every *1000* years. About 1800 the sum total of human knowledge began to increase at a steadily more rapid rate.

Now, in this seventh decade of the twentieth century, it is claimed that the vast accumulation of human knowledge is doubling every *10* years. During the 1970's the rate will continue to increase exponentially and the vast store of human knowledge will double every *5* years.

No living man can hope to encompass the span of human knowledge within his lifetime. This is the kind of world the mind of man is creating.

The increase in knowledge has been accompanied by vast social changes. Institutions, organizations, governments, religions have changed. The whole world is a state of flux and upheaval. Obsolescence of machines, institutions, inventions occurs constantly. Man seems to be caught in a continuous struggle to remain contemporary—to avoid having his ideas, concepts, understandings and knowledge become obsolescent, not occasionally during his lifetime, but continuously all the years of his life.

We constantly cope with change. And we are in mortal danger of discovering that our techniques and approaches are outdated, and are losing their effectiveness. Sometimes, in our efforts to deal with change, we forget that change is not accidental, change is not automatic, change is not inevitable! Someone or something always starts change; initiates change, always keeps it moving.

What is the role of education during this era of revolutionary change? Does education accept a passive role, and respond to the impact of change? Does it attempt to equip man to deal with change? Or does education adopt an aggressive role and seek, foster and create change?

Is it true that man has created all these tools and machines and devices of technology, only to discover that he has become their slave, not their master?

Has man, himself, changed during the ages, or has he succeeded only in changing his environment, in creating this dizzy schizophrenic external world? Is man capable of changing himself during this era of supremely accelerated change?

Is it true than man once thought himself a child of God, but now finds himself but an ex-ape, and seems destined to become an industrial animal?

If man ever learns the answers to these questions it will be only by means of education—education that continues during all of each man's lifetime, and during all the life span of the human race.

Education—continuing education—is supremely essential in this twentieth century, not only because "there are no islands any more," but more because "NO MAN IS AN ISLAND," as John Donne wrote many, many years ago.

CHAPTER II

TOMORROW

A Unique Quality of Continuing Education

The term continuing education possesses a fascinating intrinsic quality. Continuing education implies "learning in progress" at the present time, and learning going on and on as long as the mind of the individual remains capable of comprehension and application.

As long as a man lives he remains an uncompleted person. We will be what we will to be is the prime motivation for continuing education. No person even approaches spiritual and intellectual completion except by a lifetime of learning.

One of the universal attributes of man is his love of travel, his deep seated appetite for adventure, which not even the snows of old age are ever quite able to quench.

Perhaps this is why Alfred Tennyson in *Ulysses* was able to describe this quality of the human creature with poetic genius and compose lines which will be quoted as long as Literature lasts.

> "Yet all experience is an arch wherethro'
> Gleams that untraveled world, whose margin fades
> Forever and forever when I move.
> How dull it is to pause, to make an end,

To rust unburnished, not to shine in use!
Some work of noble note may yet be done,
Not unbecoming men that strove with gods.

"The lights begin to twinkle from the rocks;
The long day wanes: the slow moon climbs: the deep
Moans round with many voices. Come, my friend,
Tis not too late to seek a newer world.
Push off, and sitting well in order smite
The sounding furrows, for my purpose holds
To sail beyond the sunset, and the baths
Of all the western stars, until I die."

Much of the history of the human race has been devoted to journeys and voyages. Families, clans, tribes and whole nations have crossed deserts and trekked over mountain ranges. Continents and oceans have been ineffective barriers to the wanderers of past ages. The literature of almost every people is filled with stories of heroes who have dared to explore the Beyond. The tales of Abraham, Prince Theseus, Leif Erickson and Marco Polo are but prologue to the sagas which someday will be written about twentieth century Russian and American astronauts. The ultimate of human adventuring–the journey to the stars which beckon to us nightly–will someday occur. How profound and how far-reaching the effects of this journey will be upon mankind can be only imagined at this time.

The most intriguing part of any journey is the choosing of a destination. The planning and the journeying are more prosaic. The destination is the basic aspect of a journey, for it determines the direction to be taken and influences the route to be followed.

In a most real sense the learner is an adventurer. The questing for knowledge, the searching for answers to the Unknown is a life-long journey for intelligent men and women.

The charting of the pathways to knowledge is a responsibility of the educator. He is truly a social and educational cartographer whose charts and road maps serve to guide the young and to counsel the mature.

The search for, and the discovery of, knowledge and truth is a boundless adventure of the human mind and spirit. This is our eternal quest. Its explorations, its discoveries, inventions and insights afford man the excitement and the exhilaration of a game—the grandest game the human race has devised. Man's grand adventure into the external world of the universe has so preempted his energies and interest that the equally rewarding adventure into man's internal world—himself—has been neglected. This, too, is an urgent challenge to continuing education.

Men and women who are young, young in years and in mind and in spirit, who persist in spite of the years in being young in mind and in spirit, must not squander precious days in ignoble and foolish protest. Their days ought to be invested in constructive leadership, especially in educational leadership.

The old world is dead. The new world is just

taking form. Wise, sound leadership is desperately needed.

"Two world wars in thirty years spelled crisis, and the atomic bombs over Japan punctuated it in a way that no one could ignore. Yet the crisis of life on earth today is not the threat of atomic catastrophe. That is only its most acute and obvious symptom. The crisis is the emergence of intelligence, and its intervention in the course of evolution.

"Is the present crisis essentially different from those of the past? Man, the product of 2,000,000,000 years of patient protoplasmic experiment, has been on earth only a moment of geological time. Yet he already holds the power to destroy his culture and even his species. A little later he may have the power to destroy all life, even the planet. Outwardly he shows little awareness of his responsibility or his peril; but the fear is seeping into his bones, and it may save him. If he accepts his responsibility and applies his intelligence to such problems before he is overcome by them, he may keep open the way to a really civilized society. In any event, the present order of society is intolerable and man will not willingly go back the way he has come."*

Tomorrow is an inexorable adventure for every person and for humanity. There is excitement in the thought of tomorrow—and hope, and optimism and good fortune. There is also risk and danger and the excitement of discovery, the lure of a possible serendipity. Tomorrow is a happy word.

The magic in the term continuing education does

*J. H. Rush, "Ten Thousand Years," *The Saturday Review.* January 25, 1958.

not lie in attempting to utilize science and technology to blueprint the future. No one can accurately chart the future. Some of the roads cannot be marked at this time. Some of the potential destinations of marching mankind will remain shrouded in the mists of the years yet to come. The magic lies in our belief that continuing education will enable us more effectively to cope with TOMORROW.

The real fascination of tomorrow lies in our belief that the years immediately ahead will make almost impossible demands upon our wisdom and skills as leaders and educators. There is a fascination to any consideration of what tomorrow might be like, and to a critical examination of the probabilities of being able to manipulate future developments for the benefit of society.

Canadians and Americans have always been tomorrow-oriented. Immigrants from other countries have come to these shores dreaming of a brighter, richer future, if not for them, for their children, and their children's children. This explains our almost fanatical trust in education. We believe education can insure for us and for our children the future we have dreamed. We can never be certain to what extent we may be able to modify or control change. Most educators believe that we *must* make the attempt.

In 1965, the Rev. Martin Luther King, Jr. stood on the steps of the Lincoln Memorial and, in words of almost lyrical eloquence, described his dream of the America of Tomorrow:

"I have a dream today . . . I have a dream that one day every valley shall be exalted, every hill and mountain shall be made low. The rough places will be made plain, and the crooked places will be made straight. And the glory of the Lord shall be revealed, and all flesh shall set it together. This is our hope. . . . With this faith we shall be able to work together, to pray together, to struggle together, to go to jail together, to stand up for freedom together, knowing that we will be free one day. . . ."

Prejudice, hatred, injustice and ugliness have been with us far too long. We cannot expect that additional laws and outlays of money will by themselves eradicate these monstrous evils from our society.

Tomorrow is our greatest enigma. It is our greatest gamble. It is our greatest and most enticing adventure. All thru tomorrow we shall need our best professional skills, the most up-to-date knowledge, all of our store of creative imagination and courage. Especially, when we face days of disappointment, failure and frustration, we will need optimism based upon the conviction that education will help win the battle against disaster.

Because—who knows—through our dedication and perseverance Martin Luther King's dream for America might become a reality. The hope and the promise of Tomorrow lies heavy on the shoulders of educators.

Chapter III

NEVER THE SAME RIVER

Foundations of Continuing Education

The Greek philosopher and pessimist Heraclitus is given credit for the aphorism "no man steps into the same river twice." Its truth is obvious. No individual continues to be physically or intellectually exactly the same, even from minute to minute. The social order is in flux and tumult continually.

It is conceded that the analogy of society and a river is more figurative than literal. The molecular components of a river ceaselessly move while the river flows onward. Society continues to change as the generations of men come and go. But the individual components of society are not passive droplets. Man is a dynamic catalyst and a creator. The river of society eternally flows onward down the endless reaches of Time, but it is a continually changing river because its components are continually changing.

Man's relationship with man, the mysteries of the universe, the reasons for his existence and the nature of the Master Scheme of Things have stimulated his thinking and goaded him into scientific and philosophic inquiry since he became capable of thought. His discoveries in science and his ontological and cosmological speculations have provided man with a variety of clues and hypotheses. These clues and hypothetical speculations have been used as the bases for many kinds of interpretations of the meaning of existence, the nature of the unknown, the purposes and ends of life. No one has been able to prove absolutely that his theories about the riddles of life are the correct ones. Each individual is free to choose the theories; the systems of thought and logic he believes are most satisfactory or valid.

Philosophy serves several useful purposes. A philosophy is far more than a moral theory or a set of ethical principles. It can be an intuitive, logical explanation of the meaning of human existence and man's relationships with the universe. Philosophy provides a perspective—a "high peak viewpoint"—from which man can look over all major assumptions, theories and contentions and attempt some orderly personal beliefs. Philosophy can be evaluative. It can assist in weakening the grip of blind tradition and encourage the individual to regard life as a creative adventure. Philosophy can provide the foundation for enduring human institutions and ideas.

Western Man has been devising philosophies ever since the Greek thinkers erected their philosophical structure which Sir Richard Livingstone described as a twin-spanned Rainbow Bridge. Their brilliant systems of thought erected the first span to bridge the chasm between barbarism and civilization. The second span which the Greeks began, but which never has been finished, reaches from apparent reality to the Ideal—from arid existence to a philosophy and purpose for living.

The ideas advanced by the questing philosophers over the centuries are not mere musty theories entombed in ancient volumes. They color, they affect, they catalyze our theories and systems of education, our forms of government, our religions, our behavior in society. Upon the foundation stones of basic beliefs and convictions we build the structures of our lives as individuals. The structures of human society—institutions; cultural, political and economic systems; education, religion, have been built over thousands of years upon the foundations of beliefs and convictions—upon the interpretations which have sprung from scientific experimentation and philosophic thought.

The significance of the contributions being made, and yet to be made by continuing education and the direction and effectiveness of its impact upon our society will be affected strongly by the philosophic foundations upon which these adult educational edifices are built.

This may be an idealistic point of view. It could be demonstrated that a fairly large fraction of American continuing education is planned and conducted by sincere individuals who have little knowledge of, or make no attempt to apply, the great philosophies to their programs. Even these individuals, it can be contended, have a philosophy of education which is the basis for the character and form of their programs. A number of pragmatic philosophies have been developed. Some are individual-centered, others are society-oriented. Much of American continuing education is so diversified that a typical program would represent a kind of eclecticism, a combination of philosophies.

That education for increased vocational and occupational competency of the individual benefits both society and the individual is the premise upon which large segments of American adult education have been developed and continues to be the raison d'etre of some of the most impressive current programs.

A variation of this philosophy calls attention to vocationally related needs which individuals and society request continuing education to serve. This idea is the basis for alien-education or education for citizenship, literacy education and remedial education for school drop-outs.

Contemporary society has given prominence to another kind of program which might be labeled therapeutic. These programs attempt to meet the needs of the lonely adult, the adult with special emotional problems, adults who have intellectual needs which education can satisfy.

The concept of social roles of the individual constituting the basis for continuing education (sometimes called the Havighurst theory) has achieved considerable popularity. This idea identifies the several roles assumed by adults during their mature years, which require education during the period these roles are assumed.

One of the tenets of American elementary and

secondary education is the belief in individual growth and development, physically, mentally, emotionally and spiritually. The philosophy of continuing education is based upon this concept—that individual growth does not cease upon graduation from formal school but continues throughout life. Continuing education attempts to project and interpret the resources of the university community of knowledge to members of the adult community.

Opportunism is an appropriate label for the philosophy which exalts individual whims and grants highest priorities to the fads and the popular interests of the moment. The criteria for program design seem to be the novel, the bizarre, the expedient. Perhaps this philosophy is responsible for some of the trivia found in certain continuing education programs which has tempted some critics to unjustifiably attack all continuing education.

Vanguardism is another philosophy which possesses a special appeal, but must not be confused with opportunism. The disciple of vanguardism is addicted to experimentation. Program emphasis is often on the untried, not because it is new, but because the experimental program might point to more effective and satisfying education of adults. The desire to experiment with new forms or new approaches is based upon a genuine attempt to be creative.

The idea that adult education should identify itself with national purposes and needs has been gaining many adherents, especially since the start of the Cold War. The Education for Public Responsibility movement is not oriented to the international conflict but promotes the desirability of better informed citizens in order to assist more effective decision making on all levels of democratic government.

The tremendous advances in anthropology, sociology and the physical sciences have given impetus to the philosophy of social imperatives. Rapid and radical social changes require adjustments in all individuals. The educators who base their programs upon societal imperatives emphasize the importance of education as adults attempt to meet urbanization, suburban culture, population changes and population mobility, technological advances, industrial developments, cybernetics and the civil rights movement.

The importance of techniques and methods in adult education is the philosophic basis of another school of educators. This group does not deprecate the importance of substance, but it points out the mediocrity of much instruction of adults, contending it is caused by ineffective use of techniques. There are some who are rebelling against the creeping standardization which has been smothering adult education. The trend to organize all adult instruction into classes and courses (always meeting in the evening!) of certain fixed numbers of sessions and hours of instruction, and relying mostly upon the lecture method, might tend to force continuing education into a lock-step of schedules and format which will sap its life and value.

The past continues to influence our values, our attitudes, our decisions, but with lessening force and impact. Continuing education, although potentially subject to the influences of the past, is reacting to the immediate demands of the present, and is becoming sensitized to the challenges of the future. There is an emerging philosophy of continuing education, the philosophy of change. The philosophy of change is rapidly becoming an accepted basic philosophy in U.S. adult education.

It is this development which has provoked many to ask: "Where is the river going? Can we discern toward what ultimate sea our society is flowing? What does this mean to man the individual? What is Man?" These questions are particularly significant for today's world, because the answers or guesses we evolve will mold all education.

Perhaps Bertrand Russell, the old English rebel and well-known philosopher, has described this idea best.

> "Has the universe any unity or purpose? Is it evolving towards some goal? Are there really laws of nature, or do we believe in them only because of our innate love of order? Is man what he seems to the astronomer, a tiny lump of carbon and water impotently crawling on a small and unimportant planet? Or is he what he appears to Hamlet? Is he perhaps both at once? Is there a way of living that is noble another that is base, or are all ways of living merely futile? If there is a way of living that is noble, in what does it consist, and how shall we achieve it? Must the good be eternal in order to deserve to be valued, or is it worth seeking even if the universe is inexorably moving towards death? Is

there such a thing as wisdom, or is what seems such merely the ultimate refinement of folly?"*

We might call this the philosophy of meaning.

Just how fundamental to adult education is a sound educational philosophy? Are we emerging into a new era, an era in which philosophy will not be considered as of seminal importance?

It may be that the emerging American adult educator will look more to science than to the historic philosophies for clues to the meaning of life; will ask more of anthropology than of theology for the bases of human values; will depend more upon psychology than pedagogy for insights about human drives and motivations; will expect of sociology as well as of history a clear perspective of man in his society.

---

*Bertrand Russell, *A History of Western Philosophy*, New York: Simon and Schuster, 1960, 14th Printing, Paperback, pp. XIII-XIV.

Chapter IV

LES MONTAGNES RUSSES DE VIE

Life's Roller Coaster

Many times during these fast-paced years of the sixties it will seem to the young adults that all of us are rushing into the future on the tremendous roller coaster of life,* propelled by forces beyond the control of any individual. He may ask himself or he may ask others—are we mere passengers in these vehicles or do we have the capability to be helmsmen of our careers and steer these craft as we desire? Must we accept life as an aleatory game in which the individual has no role as chooser of significant alternatives?

The act of engagement in human society occurs much earlier than it did a generation or two ago. Young adulthood often begins before biological and intellectual maturity arrives. Individuals no longer are content to remain mere observers of society until they reach their fourth decade. This may be part of the impact of science and technology upon education and upon society. Whatever the causes of engagement by young adults in society, the results are evident and, to many people, disturbing.

---

*The French people have given the picturesque label "The Russian Mountains" to the device known in the United States as the "roller coaster."

We are living in a period of protest against the established order and a rejection of many traditional beliefs and customs. Action has displaced belief—action without thought out purposes and outcomes. There is widespread disenchantment with frustrations, delay, red tape, stuffy procedures and review of precedents. "Yesterday" is downgraded or ignored. "Today" is now paramount. In the sense that Tomorrow meant heaven, it, too, is derided. In the sense that it means fulfilment, it ranks very close to Today.

The youthful adult seems to have acquired a distrust of many of the values and standards of earlier generations. He may ignore some of the accepted "lodestars," but he remains concerned with discovering reliable criteria by which to evaluate the traditional lodestars. His concern is more about the process of valuing and less about values.

Eventually the young adult may realize that he possesses the ability to respond to the challenges of life. He can persist in trying to make more than a dice game of his life. Perhaps before too many years he learns that the role of "cybernaut" or steersman of his life is one of the most significant roles he must assume.

The young adult is attracted by the philosophy of futurism. Some persons have described futurism as the credo of a new cult of American corporation executives, scientists, technologists, educators and governmental careerists. It is far, far more. Futurism is a significant quality of leadership forged in the turbulence and competition of the contemporary world. Futurism is firmly established on the conviction that man has not lost his capacity to foresee and to forestall; that he will NOT destroy himself nor his earth. But disaster looms if the young adult embraces futurism as a shallow Pollyanna attitude toward the deeply rooted ills and problems of contemporary society.

The early involvement of young adults in the affairs and concerns of the community, of the university, of the nation, of all humanity may be the most hopeful development of this century.

Certain processes and actions have been set in motion in the factories and laboratories of this age, which if not interrupted or reversed may prevent this planet from remaining the home of man. Pollution, neglect, biological or chemical warfare may result in destruction of our natural environment. If this happens it will not be useful to perpetuate ideals and values. They, too, will perish.

Today's society is being transformed by the growth of vast sprawling ugly cities. Will the result be more evil, more degradation, more hatred, more decay in law and order? Or, will earth become a better more beautiful place because of the new cities we build?

Scientists are learning to do things to the human genes, and to manipulate the human psyche. No one can predict what discoveries will be made in the laboratories of the world or what their potentially awesome effect upon mankind will be.

These three phenomena—violation of natural environment, urbanization of society and discovery of the nature of life—confront mankind with the most terrible dilemmas of all history.

It is well that our able thoughtful young adults are not procrastinating in their assumption of adult responsibilities in society. Hope for a better future will increase in proportion to the speed and effectiveness and vision exhibited by the new generation as they mount the monster Roller Coaster of twentieth century existence.

This early sharing of responsible leadership and power by the young adults, with a more experienced generation, demands more knowledge and greater skills than most of them have acquired during their years in the schools and universities. If the hopes engendered by intelligent engagement of young adults in the problems and dilemmas of society are to become realities, continuing education will have to become more attractive and more helpful to this generation than it has yet succeeded in doing. The challenge of the tasks they are asking their elders to participate in will require of them purposeful continuing learning on a level and of a quality not achieved during their adolescent years.

Perhaps experienced administrators of continuing education should welcome numbers of youthful, experimentally-minded, resilient, stubborn young men and women to their ranks, to share in devising and designing 1970 models of continuing education for young American adults.

CHAPTER V

WHY BITTER TEA AT SUNSET?

Continuing Education and the Retiree

Few phrases are more melancholy than "bitter tea at sunset." All the terrible loneliness, the hopeless alienation, the empty endless days that fill the existence of many retired men and women are epitomized in these four words. This is the fate to which communities, corporations and society doom their educators and scholars, their leaders and executives, their professional men and women.

Perhaps John Ciardi had this in mind when he wrote, "the advantages of dying old are greatly exaggerated." Surely a dynamic, resourceful and enlightened society can provide experiences for men and women who have attained the seventh decade of life—experiences which are rewarding to the individual and enriching for the community.

The sunset years of human life should be sweet not bitter, happy not filled with fears and tears. These years should not be futile years in which intellectual vigor, special talents and creativity are allowed to decay because of disuse. It is an inexcusable waste to force able men and women to live out the remaining years of life in wearisome repetition of shuffleboard, enduring the endless boredom of card games until mercifully terminated by the visit of the undertaker.

Old age could be the greatest opportunity for continuing education. There are alternatives to "bitter tea at sunset."

There are entire mountain ranges of accumulated human knowledge, in scores of areas and categories, waiting to be learned and mastered. The fast moving years of adolescence and the work-packed years of middle age provide too few opportunities to learn and understand this vast store of knowledge.

Even more enticing to some are the vast continents of undiscovered and unexplored knowledge, waiting to be researched and probed and understood. There are songs to be written, symphonies to be composed, works of art to be painted or sculpted, books to be written.

There are endless things of beauty and wonder to enjoy and appreciate and experience. There are vacant posts of leadership and consultantship waiting for persons with wisdom and experience and seasoned judgment. Howard McClusky has often remarked, "there are some tricks that only old dogs can learn."

If there is a dearth of imaginative resourceful educational leadership in communities wherein live retired, unused, unlearning, unhappy individuals, let these individuals assume the initiative and plan together for the kind of programs they desire. This is one of the productive activities in which the wisdom and experience of retirees can function superbly.

It is practically possible for a group of retired men and women to plan and work together to launch and operate cooperatively a "Free University" or a "People's College." Utilizing the resources available within such a group there should be little need for large financial outlays.

The retired leaders, decision makers, long-range planners and thinkers, more than anything else miss the accustomed exercise of their capabilities. The possibility of inventing new and much needed new ventures in continuing education should provide these men and women with almost unlimited opportunities for the exercise of imaginative leadership, the application of wisdom and the utilization of experience.

Among retired Americans there should be the manpower, the brain-power and the spirit-power to man the kind of uncommon education envisioned by Henry Thoreau in *Walden.*

> ". . . It is time that we had uncommon schools, that we did not leave off our education when we begin to be men and women. It is time that villages were universities, and their older inhabitants the fellows of universities, with leisure—if they are indeed so well off—to pursue liberal studies the rest of their lives. Shall the world be confined to one Oxford forever? . . . New England can hire all the wise men in the world to come and teach her, and board them round the while, and not be provincial at all. That is the uncommon school we want. Instead of noblemen, let us have noble villages of men. If it is necessary, omit one bridge over the river, go round a little there, and throw one arch at least over the darker gulf of ignorance which surrounds us."

If our distraught society is too occupied with urgent problems of these turbulent times to provide its retired citizens with special educational services and programs, then these retired individuals, who are still active citizens of the nation and members of society, have a clear obligation. It is their duty to exercise the traditional American perogative of ingenuity and self-reliance. Using their own resources they should provide the programs and activities which someday the nation might provide.

Chapter VI

FEW EARTHLY THINGS MORE SPLENDID—

University Sponsored Continuing Education

This title is abstracted from an often quoted paragraph of John Masefield:

> "Few earthly things are more splendid than a university—"

It is not the university which in its contemporary stages of development may or may not be entirely "splendid," but the fundamental purposes and functions of the university that we will discuss.

An endless river of young men and women pours thru the gates of American schools and colleges every year. One and one-half million high school diplomas will be awarded in 1969. The assembly lines in American colleges and universities this year will establish new records in mass production of a hundred thousand and more degree holders.

Learning is the chief "business" of the university. Learning by being taught, learning by experimentation and research, learning by reading and exchange of ideas in seminars; whatever the method, learning is the business of the university.

An educated person is the purpose, the end, the

goal of learning. Learning is the means not the end—the road not the destination—the process not the objective. To substitute the means for the end, the road for the destination, the process for the objective is a mistake. When educators forget or ignore this principle, learning becomes sterile, mechanical and self-defeating.

Educators might well ponder the words of Robert Redfield:

> "The experience that results in education is a successful adventure with difficulties on the way. Impelled by the wonder and puzzle of things, the mind and spirit start out toward destinations not clearly seen. But soon the way is temporarily blocked. Across it lie ideas, facts, possibilities, and implications that demand to be dealt with before the traveler may go on.
>
> "But the outcome, if it be education, is not so much destruction as creation. That which lay across the path has made the traveler grow, and he proceeds, now taller than he was."*

Redfield thinks of education as the indispensable factor needed by man if he is to succeed in the great adventure—the quest for life's meaning. But, education and learning are not necessarily or automatically synonomous.

Dr. John F. Olson, Vice President of Syracuse University, tells a delightful little story about a

---

*Robert Redfield, "The Educational Experience," *State University of Iowa Bulletin #779*. A Digest of 1954 Fund for Adult Education Lectures, Iowa City: State University of Iowa, 1961. Copyright: American Foundation for Continuing Education, Chicago, 1955.

woodsman from the far north country, who never traveled away from the hills and valleys of his home. He saved his wages carefully, and in his middle years made a long anticipated trip to the bustling big city far to the south. As he clomped along the sidewalks of the metropolis in his high boots and mackinaw, he was quite overwhelmed with the roaring street traffic, the crowds of hurrying people, the high buildings and, as Dr. Olson describes it, "the wonderful visions of contemporary femininity." Finally, he came to the main entrance of a large hotel. He stood for some time watching people pass through the revolving doors. A salesman went in, and a moment later an entertainer in a tuxedo came out. A middle-aged banker entered, and an air force colonel emerged. A cleaning lady went in, and a few seconds later a handsomely attired woman came out. A uniformed chauffeur entered and soon a nattily attired college student came into the street. As he watched, his eyes grew large with amazement. Finally, he looked himself over, picked a burr from his sleeve, sidled up to the resplendent doorman and asked, "How much does it cost to go through that machine and do you get your choice of what you will be?"

This story no doubt is fiction, but it conveys a very real message. We actually possess a device which helps us become the kind of person we desire. It is not a magic revolving door, nor an electronically operated gadget. *We label it education.* It is not instantaneous. In fact, it requires a lifetime; but the end result is certain if we persist. This is one basic reason why education is so essential.

Education is not an idea, such as freedom or immortality; but it is based upon an idea, and it is concerned with ideas. Education is not a device, such as television, the computer or the printing press; but it utilizes devices and inventions. It is not a method, like discussion, or a lecture; but it utilizes a variety of methods of instruction. It is not a discipline, but it includes in its substantive spectrum most academic disciplines and areas of knowledge. Education is a social dynamic—a human enterprise, involving many diverse institutions, agencies and organizations, serving many groups and levels of persons, having many varied forms and character, possessing numerous and continually changing purposes and goals. Education is a vital human activity, based upon a fundamental characteristic of human existence.

Olson thinks of education as a process which permits individuals to become the kind of person they really desire to be. To human beings, *becoming* is of greater consequence than *being;* and to *become* is a far greater achievement than just *to be.* This is the seminal purpose, the prime challenge of education.

If learning is the means, then the learner is the raison d'etre for all of education. If learning is the "adventure" then the individual is the "adventurer." All of man's carefully constructed paraphernalia of learning aids and devices and programs are but charts and road maps for the use of the "traveler."

If education is indispensable in seeking the meaning of life, if education is the only process whereby an individual can achieve this maximum potential, then education must be a continuous life-long adventure. The university, if it is to honor its deepest obligations to society and to individuals, must regard on campus education of adolescents and young adults only as a necessary preliminary to the vastly more significant obligation of life-long education of these youthful individuals.

The learner soon discovers that he is caught up with millions of other men and women in a never ending race with knowledge.

The gnawing fear of falling behind in the race with obsolescence is as a menacing cloud looming over the heads of men and women in the professions.

This situation contributes to the danger of learners accepting the contemporary attitude—continuing education is the fashion of the hour—everybody is continuing their education. A "continuing education production line" operated by an institution is not impossible or even implausible! To learn, just because it is the fashion, the accepted thing to do, is to invite captivity by the forms and procedures—by the externals of education.

Education is intended to free the mind from the shackles of blind conformity and from the bonds of fear and ignorance. The "unthinking learner" (is there such a person?) might find himself on a treadmill, continuously pursuing the forms of learning but never quite realizing the enjoyment of the substance.

Is it possible that the very complexity and vastness of the exponential expansion of human knowledge will encourage the development of pre-planned and pre-fabricated systems of knowledge for the various occupational groupings? If individuals should ever surrender control of their continuing education to any institution or organization or agency or party or government, then we can expect the arrival of some form of Utopia. It may not be "1984," or "Walden II," or a "Brave New World," but it will be a society in which the freedom to choose what is to be learned will inevitably be taken from us. The next step will be loss of the right to dissent. The learner must always be in command of what he wishes to learn. Not even dedicated educators should attempt to usurp this prerogative.

There is another danger. No profession is without its charlatans and quacks. Education is no exception. There are fads and alluringly packaged techniques advertised to be the panacea for all of the ills of society, which afflict dedicated educators and entrap many of the eager but unwary students.

We hear much these days about the potential of the human mind as contrasted with its usual performance. It is this point of view which creates our reliance upon grades and marks, examinations, intelligence tests, aptitude tests, etc., in institutionalized learning.

Michael Young, a British author, coined the word "meritocracy" to describe our society—the society which assigns status on the basis of intellectual performance instead of family or color or religion or sex or wealth. If true, it may not be entirely a happy development.

Teilhard Chardin and Loren Eiseley, two twentieth century men who uniquely combine the qualities of scientist and mystic, have challenged us to consider the evolving and developing human mind, evolving toward some far distant culmination of power and achievement. They maintain that the human mind is significant, but only in terms of its ultimate development. The human mind has progressed far beyond being only a sensation recorder, a chemical-electro stimulus response, a musclemover. It must continue its progress in creative achievement.

This is a somewhat intriguing argument for stretching our contemporary minds to their limits. It raises some very interesting questions about the extent of our responsibilities in the continuing creation of mankind.

Educators must become convinced that continuing education will never achieve its most significant role in society until it accepts the imperative of utilizing education to improve the quality of human life.

We desperately need a whole new breed of continuing educators: leaders who believe in the new role and imperative and are committed to devoting thought, time and energy to its realization.

This is not a plea for idealism instead of pragmatic administration. It is a plea for adding a new and vital objective or reason for being to the thrust of the adult education movement in North America—another dimension to continuing education as a service to society.

What is meant by the quality of human living? Perhaps it can be described by quoting from Robert J. Blakeley:

> "Our being is a perpetual perishing and a boundless becoming. At times we catch a glimpse; is it starlight reflected on the water, or the image of eternity? In a measure we can choose and realize what we will to be. . . . We don't know that there are human beings on other planets. We do know that there are human beings on ours. . . . Might we exchange trivia with alien forms of life a hundred light years away before we communicate with our own kind across the tracks, across the creeks, and across the barbed wire fence? We will sweep all the frequencies of the sky. Will we harken to the counsels of our hearts? We will puzzle endlessly over static from space. Will we ponder the promptings of our poets and heed the insights of our saints?"*

It is not at all certain that this added dimension to the "curriculum" of continuing education can be implemented only by university courses, conferences, experimental group studies or programmed learning.

It might be achieved if all institutions and agencies currently active in continuing education in the U.S. embrace this goal. Its accomplishment would require a radical overhaul of a vast segment of the continuing

---

*Robert J. Blakely, "The Next Human Nature," *Adult Leadership*, Vol. IX, No. 7, Chicago: Adult Education Association, January, 1961, pp. 204-208, 217-220.

education enterprise in this country, as currently operated by colleges and universities.

Too often continuing educators relegate basic concepts to a minor place among the concerns which preempt their interest, energy and enthusiasm. One of the basic concepts which is absolutely fundamental is the function and obligation of the contemporary American university in continuing education.

A continuing education prophet of an earlier generation, Lyman Bryson, addressed a special convocation of the University of Michigan in 1937, on the occasion of the celebration of the university's centennial. His remarks concerning the continuing education function of a university are as vital and as pertinent now as they were in 1937. Some of his statements were as follows:

"We are not facing reality, however, unless we realize that something more will be needed than a continuation of past success if the universities of tomorrow are to be leaders in the adult education movement. If the universities are to continue to be the capstone of the educational system they must somehow manage to get and hold leadership in the voluntary educational activities of mature men and women. They may decline to do so. The universities are free to neglect their opportunities as well as to embrace them. We believe in freedom, but freedom involves responsibility to take the consequences of an act, so if the universities of today do not become leaders in the adult education of tomorrow, their importance in the educational system will diminish.

"A university is a group of men and women seeking

and declaring what they believe is truth. Knowledge must have a house, and in these days we must have also laboratories and books. But the halls and equipment are secondary. The only thing that makes a university hall any more important than a warehouse is the thinking and teaching it has sheltered. And can this, the true spirit of the university, be made the possession of the whole population? That is a question to which we do not know the answer. But we must find the answer, and the answer must be as generous as we can make it.

"These two questions are, I believe, most important in considering the place of the university in continuing education. How can the university serve all the people? And how can it produce men and women among its own students who will not stop using their minds, but will serve the people wherever they are with the best that is in them?

"It is no small or timid role that a great university can plan in such a drama. But universities believe in tradition. The oldest tradition in human history is the tradition of the fight for the truth. Other traditions, against which this one incessantly contends, are born and die. This tradition of giving new ideas a fair judgment was born in the first beings who could be called human, because all progress and civilization have been its creation, and it still lives.

"By such action, and by all the devices which we now know, the university can make its spirit a part of the life of the whole commonwealth. It can teach the young to use their critical powers, not as playthings in an irresponsible monastic safety, but all through their working days. It can keep in touch with them, not only by an emotional loyalty, but still more by a continuing inspiration to thoughtful boldness. It can offer to the society of which it is a part the whole of its resources, its men, its books, its oldest learning and its newest discoveries, and it can humbly study the ways by which those things can be made useful to every kind of human need.

"A great university will uphold those who dare to think, inside its walls and out. They will be few enough, but they are the leaven. A great university will care most, not for the driven hordes that find spiritual rest, as of death, in this ism or that, but for a society of men and women who hold to reason, to science, and to the difficult life of freedom. The world is now full of men who deny that reason has any meaning to them or to the state. I would rather see the house of this university levelled to the ground and its scholars put to useful, honest day labor than to see it become the servant of any tyrannous state or intolerant people. These things are easy to say; hard to do. But a university is a gathering of fine minds and stout hearts. What other claim can it have on our loyalty and our faith?"*

The unending task of the continuing educator is choosing between essentials and non-essentials, between the things which are important and those of lesser consequence. He tirelessly must prevent time, energy and leadership from becoming dissipated into the dry channels that lead into the desert of non-productivity. All this requires unusual insights and

---

*A University Between Two Centuries; the Proceedings of the 1937 Celebration of the University of Michigan, edited by Wilfred B. Shaw. "*Continuing Education*" by Lyman Bryson, Ann Arbor: University of Michigan Press, 1937.

capabilities. Being a continuing educator is neither easy nor insignificant.

This is one reason why administrators of continuing education must persist in continuing *their* education. They must be discerning individuals. In the following example, Mr. Dooley cautions Hennessy on the difference between reading and thinking.

> "Readin', my friend, is talked about by all readin' people as though it was th' only thing that makes a man betther thin his neighbors. But th' truth is that readin' is th' next best thing his side iv goin' to bed f'r restin' the mind . . . . Believe me, Hinnissy, readin' is not thinking'."

University educators, especially those directly charged with leadership and administrative responsibilities, cannot afford to slumber through the current social revolution. There must be a rigorous examination of some of the philosophical barnacles which are obstructing the ability of universities to develop bold new programs with alacrity and confidence. The time-worn shibboleth "appropriate for a university" must be reinterpreted within the perspective of human needs.

Perhaps the university educator could accept the following statement, written as a paraphrase of a paragraph of the acceptance speech of William Faulkner on the occasion of the award of the Nobel Prize, in Stockholm, in 1950.

> "It is the *educator's* privilege to help man endure by lifting his heart, by reminding him of the courage and

honor and hope and pride and compassion and pity and sacrifice which have been the glory of his past and which can make possible an even more glorious future. The educator's voice need not merely be the record of man, it can be one of the props, the pillars to help him endure and prevail."

Yes, it is possible that in a very few years there will be many continuing education programs and services, developed by American universities, which will provide evidence for the validity of John Masefield's words:

"Few earthly things are more splendid than a university—"

CHAPTER VII

CHILD OF THE UNIVERSE

Continuing Education and the Individual

An appropriate introduction for a discussion about the centrality of the individual in the totality of education comes from a paragraph written several years ago by Roger W. Riis.

> "I admire the human race. I do indeed . . . . We have done and are doing a better job than anyone has any right to expect . . . . Of a persistance, a daring, and ingenuity impossible to surpass, we find ways to move easily under the water and through the air. Now we speculatively eye our neighboring planets. It should astound no one if man one day begins to move among these planets . . . . Daunted by nothing, his horizons constantly recede, the territories of his possession and use expand and expand. Whenever he comes to an impassible obstacle, an apparently final barrier, he goes to work at it, and in due time surpasses it . . . . I do not think he has limits. I think he is a child of the universe who inherits eternity."*

Answers to the age-old query "What is Man?" have been attempted by poets, philosophers, physical scientists, anthropologists, sociologists, psychologists

*Roger William Riis, *I Admire the Human Race*, New York: Book-of-the Month Club, Inc., 1951.

and biologists. All answers are of interest to educators, of course, who also often attempt to answer this question. The answers which educators accept are important because they determine the major purposes, establish program priorities and primacies, and pervade the methodology of all education, including adult education.

All of earth's societies do not accept the same point of view about the significance of the individual. Much has been written in Western Society about the importance of the individual. Much has been said about the centrality and the value of the individual in a democracy—that he and his welfare outrank the group, the community, the state, the organization and the institution.

The implications of this belief are many and pervasive. Proud buildings, complex staff organizations, technological installations are but means, not ends. The panoply, the gadgets, Madison Avenue approaches are appendages to the major effort. Even the financial grants of foundations and industries are to be regarded as expediting and facilitating forces and should not be allowed to become the determiners of policy and purpose and procedure. The ultimate objective of all education might be described thus: to assist each individual to achieve his fullest potential and effectiveness as a human being, thus increasing his effectiveness in the roles and tasks he assumes as a member of society, enhancing his general welfare, satisfying his hunger for knowledge and creative achievements, and increasing his happiness and satisfactions.

Some observers and students of contemporary life are not always agreed that this idealistic relationship of man to society actually exists even in democratic nations. Contemporary writers paint stark pictures of the individual enmeshed in the cruel gears of the social machine.

Whether we lean toward the idealistic or toward the realistic version of the relation of the individual to society, the inescapable fact remains, we can only educate the individual; we can only improve society by improving individuals.

There are many concepts of the nature of education. Just as our convictions about the place and importance of the individual determine the central prime purpose of all education, so our concepts of education catalytically affect the very essence and character of the educational process.

In a strange, perhaps almost poetic, yet very real and significant sense, education and creation are synonymous. True, this is not indicated in the thesaurus or the dictionary. However, there is much to indicate the validity of this contention.

Man was not "created." The creation, and the education, of man began in one, or perhaps several, of earth's warm valleys thousands of centuries ago. It has been continuing ever since, and will continue until the end of Man knells the end of Time. Within these uncountable millenniums the brief spans of all men begin and end, repeating endlessly the processes of creation and education which continue from the moment of conception to the moment of death.

This duality, this dynamic interaction between the two aspects of the development and maturation of man was a dominating occupation of the human race in the era of tree huts and cave dwellings, in the times of mud villages and river cities. It continues thus in these times of skyscrapers and split-level houses. Men are occupied with:

the educating and the creating
the learning and the teaching
the seeking and the imparting
the questing and the reporting
the thinking and the doing
the studying and the making
the testing and the using
the planning and the building
the educating and the creating

If we accept the viewpoint that education is the process of creation—of individuals and of society—then much of the dust and fog about aims and goals, about tasks and responsibilities, about methods and techniques dissipates.

It is important that educators understand the unending nature of the education of an individual. Some psychologists and psychiatrists include pre-natal influences as part of the educative process. All of us would include early childhood, pre-school years, elementary school, secondary school, the college experiences or vocational preparation and the multiform experiences of adulthood. These experiences cause and help mold the characteristics of personality, the bases of human behavior and establish many of the factors of the drives, compulsions and motivations of the individual. In each developing episode desires are cultivated, needs created, interests and tastes developed. These cumulatively affect the future episodes of the individual's life. Points of view and opinions change to biases and prejudices. The cumulative effect of learning and experience during the years of life render the task of educating adults more complex, if not more difficult than the educating of children.

The continuing education of individuals is an enduring factor in the progress of society. Its cumulative impact upon millions of individuals actually assumes the characteristics of a kind of continuing education of and for society. Acceptance of this point of view by the educator may assist in destroying any mental conflict of purpose (education for the individual versus education for society); but, at the same time it may provide him with assurance of dedication to man's highest aspirations.

This suggests a dedication of educational institutions to the development of the individual—not for a few short years—but a contributing to his development and education during all the years of his intellectual competency.

Whatever is the relationship of the individual to society, he can never entirely divorce himself of his obligations to himself. His obligations embrace a range of matters—moral, educational, vocational, health, religion, personal appearance, relationships with others. In the abstract, we can think and talk and write about the centrality of the individual in adult education. Actually, the individual has certain responsibilities to be discharged, which include his responsibility for educating *himself*. The individual adult is not helpless. He may suffer certain handicaps, psychological and physical. But he owes much to himself. In discharge of that obligation continuing education is the key. Much of continuing education can, and ought to be, "self-centered."

What then is the most productive relationship between continuing education and the individual? Is the individual to be a passive receiver of fabricated programs and services? Or, is the individual to assume the role of an active, demanding, participating partner in the process of continuing education?

These are questions of fundamental importance. How they are answered determines the appeal of continuing education to adults, the character and quality of instruction and the effectiveness of continuing education serving the needs of adults.

The most insulting label used to describe adults engaged in continuing education is "enrollee." The term "student" is banal. Meaningful words such as "participant," "reactor" or "contributor" are more desirable. Adult participants in continuing education should function as co-designers, co-learners, co-leaders and co-instructors. *The continuing education seminar or class in an alembic in which the participants are catalyzers of learning.*

The answers do not lie entirely in the province of the technical operator of programs. Nor can these issues be resolved only by changes in procedural policies. The vision and commitment of continuing educators determine substantially the relationship between the individual and continuing education.

Another significant factor in determining the quality and character of continuing education is found in the spirit and attitude of the individual. Continuing education is only in part an affair of the intellect. It is also of concern to the heart. What Pericles said in eulogy of the heroes of Athens is pertinent.

> "Knowing that the secret of happiness was freedom, and the secret of freedom a brave heart, they did not stand aside."

Chapter VIII

THE FUTURE TENSE

Social Imperatives of Continuing Education

The sometimes cryptic, sometimes poetic philosopher Soren Kierkegaard observed man's concern with the unknown future in the somewhat enigmatic lines:

> "He who fights the future
> has a dangerous enemy.
> Through the eternal
> we shall conquer the future."

Man always has been oriented toward the past, the present and the future; but it is to the future that the compass of his spirit is most sensitive. The extent of man's reliance upon education indicates the degree to which he is attuned to the future, for the essence of education is preparation for the future. Yet, it seems strangely incongruous that in our American society, enriched and embellished by brilliant technological achievements, there should be so much foreboding and fear about the future.

Contemporary American society is vastly different from nineteenth century society, almost unbelievably different from fifteenth century European society. Society of today is fluid, and offers an observer a kaleidoscope of changing characteristics. It has been variously described as Affluent Society, Mass Society, Corporate Society, Dynamic Society, Expert Society, Free Society, Democratic Society, Welfare Society—even "Waist-High Society."

Society is man's largest, most inclusive, most persistent community. It is a total community embracing human beings, their social inventions, devices and institutions. Education is one of the constituents of the community of society. Education is the creation of society. Society regards education as its trustee and holds it responsible for its continuance and its welfare. But education is also a dynamic, a leaven, a social catalyzer; so society is continually experiencing a thrust toward change at the same time it is being encouraged to cherish the traditions and institutions of the past. In this sense education has in itself become one of the major dynamics of American society. Neither politics, religion nor economic affairs equal the sustained interest and concern of our society in education.

Another factor which supplies much of the power behind the expansion of continuing education is the exponential expansion of knowledge. One result of this "knowledge explosion" is the phenomenal advance of technology which has triggered a chain reaction of previously unequaled revolutionary developments in our society. Automation in industry and manufacturing has set in motion a series of social reactions which permeate our entire social structure. The wise use of vastly increased leisure is one example of the many effects of this. The advent of space exploration and the discovery of nuclear energy have been too recent to evaluate, but their ultimate results without any doubt will transform our world and its educational needs. Modern advances in transportation have made neighborhoods of entire nations. New developments in communication technology have made instantaneous worldwide transmission of events and of ideas possible, and have also increased the danger of mass stereotyping of thought and opinion. It may be that cybernation is setting in motion the greatest transformation in society since the Industrial Revolution.

Tremendous changes in the size and makeup of earth's population constitute another societal dynamic. The advent of the megalopolis, those tremendous sprawling urban areas, has created a new urban culture and triggered a progressive dislocation of countless thousands of families. Urbanization is a social-economic syndrome which has transformed the lives of millions who live in the teeming cities. Often overlooked is the pervasive effect upon millions who do not live in urban areas, but are profoundly affected by urbanization. A breakdown of primary groups seems in progress, complicated by a tremendous increase in individual interdependence and specialization. The removal of the decision making process from small intimate group relationships to remote impersonal agencies signals another basic change. "The Lonely Crowd" is more than a title of a best seller, it is a very accurate description of the plight of the individual in current urban society.

Fear of extinction and fear of loss of our traditional freedoms insidiously pervades our daily existence. Values and standards are often eroded. Millions increase their reliance upon religion and at the same time millions become apathetic toward this traditional source of assurance and faith.

Natural and social scientists have long been worried about the widespread lack of understanding about man's environment and its effects upon the lives of individuals and upon our entire social structure. Few individuals understand what environmental changes will do to the future of the human race.

Pollution of the earth's atmosphere is increasing. Smoke and chemical fumes from industry and exhaust gases from millions of automobiles have been poured into the air for many years. Now we are faced with a development of even greater danger, the difilement of the earth's atmosphere by the by-products of nuclear explosions.

Pollution of the earth's water supply continues apace in spite of sporadic and spasmodic attempts to improve the situation. Rivers, streams and lakes, once pure and sparkling, are no longer the habitat of living creatures or a source of enjoyment by human beings. Accompanying water pollution is a tremendous increase in the consumption of water by America's industry. The cumulative effect of water use and water pollution is a frightening shortage of water, especially frightening to those who see its effects upon further growth of our population and upon expansion of industry.

Many scientists are concerned about the speed with which we use up our resources of ores, oil, and other minerals. We turn to other nations and other continents; we contemplate the eventual use of atomic energy. But to unborn generations we will bequeath an environment which has been tremendously changed by our exploitation of natural resources.

The rapidly disappearing wilderness in America has been partially adjusted by our state and natural parks. Apart from their nostalgic appeal, or their recreational uses, little has been said about the contribution of wilderness areas to maintaining watersheds, flood prevention and conservation of animal life.

Man is the only creature who persists in desecrating the wilderness and the beauty spots of the world. If left unchecked and uncontrolled our Grand Canyons and Yosemite Valleys would become clogged with bottles, tin cans and waste. Man threatens to bury himself under mountains of his own garbage.

Still another societal imperative grows out of the contemporary revolution in human values. Human behavior is not the result of any one factor, but ethical and moral standards are a very powerful determinant. Educators are concerned about the changes in public and private codes and standards because of their effect upon the very character and quality of society, and the worthwhileness of our existence.

Certainly these four or five imperatives do not

constitute all the needs and issues in our society. These, and others of varying urgency and importance confront educators with unparalleled challenges as men and women ponder their contemporary dilemmas and the terribly important choices they must make from the alternatives available to them.

Citizens may ask of educators three pertinent questions:

> What can education do to help us solve these tremendous problems? What hope dare we have of finding solutions in this generation? Can society survive this inundation of critical problems?

CHAPTER IX

THIS GREATNESS

Continuing Education and Our National Ethos

The real character of a nation—its heart, its mind, its soul—is seldom expressed in the fleeting superficialities of the day. It is revealed in times of testing, in hours of travail, in how its people rise up to meet the crises of time and circumstance. It achieves visible form in its institutions. It can be seen reflected as in a mirror in the writings of its poets, philosophers and statesmen. It assumes substance and reality in the ideals and convictions of its citizens and in their national goals and aspirations.

The American Ethos, Americans like to think, is unique. The character of our nation has been shaped in the crucible of four centuries of exploration, sacrifice, courage and inventiveness. Many peoples and races have contributed richly. Americans are an amalgam of the sons and daughters of many nations; a fusion not yet perfect, of the hopes, the traditions and customs, the biological inheritance and the cultural legacies of many peoples.

Education has been woven into the fabric of our national life since colonial times. Academies and universities were founded early in the history of the states, and the westward march of the nation across the continent was accompanied by the cadence of founding schools and colleges. Few nations have reposed so much confidence in education, or have relied so completely upon education as insurance for our future well being.

Continuing education is not a tardy and belated contributor to our national life. Planned education of adults has been underway since very early in the nineteenth century. Yale and Harvard universities presented lectures to the public before the Civil War. In 1824, Josiah Holbrook started the famous Lyceum movement. Fifty years later Rev. John A. Vincent began the Chautauqua which was to spread into almost every community in the nation during the next generation. Early in the twentieth century President William Rainey Harper of the University of Chicago and President Charles R. Van Hise of the University of Wisconsin shared in the beginning of university extension education in the United States. The influx of foreign born was the challenge which created public school sponsored Americanization classes and alien education. Other developments such as the Land-Grant University, the rise of the evening college, public school adult education and labor education programs have added to the expanding dimensions of what some call "the adult education movement."

Now the world is well along in the second half of the twentieth century and the 200th birthday of the United States is growing so near that preparations are being made for its observance. What is the role and the responsibility of continuing education to our nation in this era? What identification of adult education with national perils, national goals and national issues is justified?

There are many issues and problems, of varying urgency and importance, which currently perplex the people of the United States. Some have their origin in our Western Society, some result from the assumption by the American nation of world leadership. If American education is identified in any significant way with the nation's goals and welfare, then these issues and problems logically become an opportunity for research, fact finding, information dispersal, instruction or special services in adult education.

In addition to the War for the Minds of Men, in which our nation has so much at stake, even a casual observer of current affairs could list a number of other critical issues in which continuing education appropriately could assume a role. These would include, as a minimum, the following clusters of problems.

The achievement of full citizenship status rights and societal opportunities by the American Negro. The problems of segregation and non-segregation and of all minority groups are nationwide in urgency and occurrence.

Conservation of our depleted natural resources is a problem of prime importance to our nation and a concern of adult education. The preservation of wilderness areas, the elimination of stream and air pollution are appropriate concerns of education. The challenge to continuing education is clear-cut. Can public education postpone or prevent man-created deserts in many areas of our country?

The metamorphosis of decision making in our government as urban areas increase and rural areas shrink has caused conflicts of interest in state government and disengagement of the average citizen from governmental affairs in urbanized communities.

The complexity of economic principles and phenomena frustrates many intelligent and otherwise informed citizens. There is need for public education about basic causes of unemployment, about increasing tax costs and about cycles of prosperity and depression. Programs of public education based upon explanation of international economic affairs are very timely.

Population growth and changes in segments of the population have produced a chain of issues and problems. How to use productively the abilities and potentialities of retired workers and how to protect the health of increasing numbers of the aged are increasing in urgency because of the larger numbers of adults who live beyond sixty years. An explosive birthrate is creating an inundation of youngsters which is swamping our schools and universities and intensifying an existing dilemma of our young men and women.

It seems inevitable that a greater responsibility will be given to continuing education to assist in preservation of our national ideals and to encourage achievement of our national purposes. To avoid chauvinism and a clap-trap kind of patriotism will be part of the task. In this task many need to join their dedicated efforts—the newspapers, the magazines, the pulpits, the radio stations, the television stations, the libraries, the labor unions, the Farm Bureaus, the Chambers of Commerce, the service clubs, the public schools and the colleges and universities. It is not a responsibility or a task to be left to any one institution or agency.

In our world of bewildering change, it is sometimes forgotten that the greatness of individuals and of nations can be traced to the beliefs, ideals, principles and values held by individuals and nations. Historians are prone to assign the wars and the peacetime achievements of man to economic, nationalistic or political causes. But how can history explain a Jefferson, a Lincoln or a Churchill? Education has a terrible responsibility—that of enabling individuals to select wisely their basic beliefs and values, to retain those of precious importance and to reject those of false ideologies and social tyrannies.

Our nation has changed vastly since 1776; and so has the world. But both have changed because people have changed. Men change the world and are changed in the process. Individuals change our reservoir of knowledge and consequently effect great changes in society in this accomplishment. The essence of education for citizenship lies in the fact that individuals can shape the nation of our future. Largely through education can our citizens insure that America will become "America the Beautiful" in every sense.

This greatness of our nation, like the greatness of Athens, is part of the past. It was reflected in the sacrifices of countless thousands of immigrants who came to our shores. It was evidenced in the bravery of the westward trekking pioneers. Millions of Americans have had a share in it.

Education will continue to have a vital role to play in our nation's future. The definition of this role and its implementation may prove to be the most exacting and one of the most rewarding achievements of American education.

CHAPTER X

THE AGE OF WONDER

Implications of Scientific Advances for Continuing Education

Thomas Carlyle in *Sartor Resartus* wrote:

> "The Man who cannot wonder, who does not habitually wonder—is but a pair of spectacles, behind which there is no Eye."

During most of his existence man has been content to explore the fringes and beaches of the vast unknown oceans of energy, matter, space, time and life. Only recently have scientists penetrated deep, far and fast into the mysteries of the universe. These ventures into physics, astronomy, chemistry, biology, anthropology and related sciences have made shambles of many traditional theories and postulates. A deluge of devices and inventions has resulted in revolutionizing our industrial processes. New and persistent questions have been raised about man and man's purpose and destiny.

The stark drama of discoveries of nuclear energy and of the sending of manned vehicles to probe the vast reaches of space has tended to focus attention upon these developments. They are tremendous achievements. Educators must continue to be aware of advances continuously being made in other sciences and the resulting situations and problems. There is an urgent need for man to give purpose and direction to science, instead of passively accepting the problems spawned by scientific experimentation and permitting a handful of scientists to make decisions which may profoundly affect the lives of millions now living and of many more millions yet to be born.

Alfred North Whitehead describes the dilemma of man in the Age of Science.

> "In the conditions of modern life the rule is absolute, the race which does not value trained intelligence is doomed. Not all your heroism, not all your social charm, not all your wit, not all your victories on land or at sea, can move back the finger of fate. Today we maintain ourselves. Tomorrow science will have moved forward yet one more step, and there will be no appeal from the judgment which will then be pronounced on the uneducated."*

It is significant that scientists do not ask, nor do they expect, that education will acquaint everyone with the intricate complexities of science. It is significant that they see in the results and effects of scientific advancements opportunities for educators and scientists to cooperate in meeting the challenges of the Age of Science.

There remains another dimension of human life

---

*Alfred North Whitehead, "Introduction," in Wallace B. Donham, *Business Adrift,* New York: McGraw-Hill Book Co., 1931, pp. viii-xix.

which science is permeating—the dimension of beliefs and values. The scientific age will have a more profound and permanent effect upon our religious beliefs and our value systems than any past discoveries, including Copernicus, Galileo and Darwin. To avoid utter chaos in basic convictions and a breakdown in social values and standards education should increase its efforts to broaden and deepen the bases of our beliefs and ethical systems.

Scientists and non-scientists alike are concerned about the ultimate purposes, uses and effects of science. This is also a prime concern of adult education. It is here appropriate to repeat the words of the American scientist Robert Oppenheimer.

> "A great discovery is a thing of beauty; and our faith—our quiet binding faith—is that knowledge is good and good in itself. It is also an instrument: it is an instrument for our successors, who will use it to probe elsewhere and more deeply: it is an instrument for technology, for the practical arts, and for man's affairs. So it is with us as scientists: and so it is with us men. We are as one—instruments and ends, discoverers and teachers, actors and observers. We understand, as we hope others understand, that there is a harmony between knowledge in the sense of science, that specialized and general knowledge which is our purpose to uncover, and the community of man. We, like all men, are among those who bring a little light to the vast unending darkness of man's life and world. For us, as for all men, change and eternity, specialization and unity, instrument and final purpose, community and indi-

vidual man, all complementary to each other, both repair and define our bonds and our freedem."*

Critics of scientists whose laboratory studies reveal the secrets of the universe sometimes accuse them of "playing God." This is probably an unjust charge. But there are inevitable consequences for all mankind to decisions already made by nuclear scientists. Even more startling discoveries by the biological scientists are imminent—discoveries of the secrets of life and the mechanics of transmission of characteristics from generation to generation. Should our society, either deliberately or by default, entrust to a handful of scientists, and/or technologists hired by our national government, the kinds of decisions which will produce consequences of transcendental importance to all men?

Can education prepare our citizens to participate intelligently in the making of decisions forced upon society by the breakthroughs of science?

---

*Robert J. Oppenheimer, *Science and the Common Understanding*, New York: Simon and Schuster, 1954.

CHAPTER XI

SEVEN-LEAGUE BOOTS

The Wide World—New Campus of Continuing Education

The surging new nations of earth since World War II have been marching toward maturity in seven-league boots.

Education of their citizens has enabled leadership of these nations to leap-frog across whole centuries of development and progress. Confident, efficient young men whose parents lived in huts of ignorance and superstition in the remote valleys of Ethiopia now pilot gleaming jet planes between the historic capitals of Europe and the cities of New Africa. Vividly garbed black men, only one life-step removed from the Stone Age, sit in the Assembly of the United Nations as equals among the representatives of the ancient proud nations of the world. It is education which must enable others in these lands, not yet so fortunate, to move forward into twentieth century society and culture. Much of this education might be supplied by American resources of men and money.

Rampant nationalism sweeping inexorably across mountains, deserts and seas has destroyed great empires and thrust into the unready hands of new leaders the scepters of power and leadership. Amid their rejoicing over sudden independence the citizens of these newborn nations are engulfed in the intricate problems of government, economics and international affairs. They are bewildered by the abrupt transition into the twentieth century.

This is a challenge to education. Among the leaders of American education there can no longer be preoccupation with American affairs, or satisfaction with pigmy-like deeds and accomplishments in serving the educational needs of citizens of other lands.

Telescoping of centuries of development into a few short years has created for American educational institutions a new and most unusual opportunity, an opportunity to assist in the education of the youth and the leaders in many nations. Some of this responsibility can be accomplished within the continental United States by a system of government-supported scholarships and private foundation grants for fellowships in American colleges, encouraging learners from other countries to study and sojourn among Americans. Much of the task must be done abroad and this entails transporting American administrators and teachers to other lands. Some institutions, including colleges and universities, have already assumed this responsibility. Whether other adult education agencies can carry their programs abroad seems, at this moment, less a question of appropriateness than of appropriations of manpower and finance.

During the ten years, 1951-1961, the number of foreign students in the United States increased from thirty thousand to fifty-three thousand. During the same period the number of U.S. students studying abroad (mostly in Europe) increased from about eight thousand to just over fifteen thousand. The numbers of both groups ought to be increased tenfold.

The foreign policy of the United States has leaned heavily since 1946 upon financial aid to other nations. Billions of dollars have been spent abroad for military, economic and educational purposes. Much of the load of providing administrative and teaching personnel for foreign educational projects has been assumed by American universities. Institutions which once proudly proclaimed "the state is our campus" now just as proudly state "the world is our campus." The eventual cumulative impact of these educational ventures upon the peoples of the world cannot now be accurately predicted, but it is certain to be profoundly significant. The Peace Corps might be regarded by future historians as one of the most significant contributions of the U.S. to world development.

It is important that Americans achieve an understanding of international developments and obtain information to guide them in influencing the foreign policy decision makers of our national government. Comparatively few adults attempt systematic study of international events and issues. They rely chiefly upon newspaper headlines, and radio and television commentators for information and pre-digested attitudes and points of view.

It is also important to bring to our American colleges thousands of future leaders of the Asian, African and Latin American nations, to exchange American educators with their counterparts in other countries and to establish United States sponsored colleges and technical institutes in foreign lands.

A more subtle but most effective educational effort would be the eliminating of race and color restrictions and discrimination where these exist on America's university campuses and in communities where universities are located.

These challenges to American education are too clear and too urgent to be denied or ignored. A number of major issues and problems must be studied by Americans; and a body of public opinion must be formed which can be used by our national leaders and officials as the foundation for formulation of national policies. In our conversations, our daily press, our radio and television, our magazines and our current books, over and over are found references to nuclear warfare, fall-out shelters, biological warfare, peaceful co-existence, accommodation with Russia, space exploration, the Cuban dilemma, United Nations and similar issues. Continuing education in the United States can do much to transform fear, frustration and futility into positive citizen participation in world affairs and in America's future.

There are two important issues which confront the United States—can peace be made permanent and secure, and can our legacy of centuries of struggle for our freedom be preserved? These are starkly realistic; they are not poetic pipe dreams or oratorical flights of rhetoric.

These two issues confront diversified and fragmented continuing education with an urgent opportunity to devise some effective procedure for joint planning of appropriate program roles. There must also be cooperation in mutual assistance among the agencies and groups participating in serious study of these issues.

To a large degree, the extent to which the resources of American education can be brought to bear to find solutions to these issues will be determined by the vision, the dedication and the resourcefulness of the leaders in continuing education. However, the responsibility and the opportunity are not to be assigned only to educational leaders. No citizen of our republic can avoid some share in a realization that we are living today in a new world, a world greatly different from the old world of a few short years ago.

Americans do not fully realize that we are but one nation, one people among many. The many peoples of earth—what a fascinating bewildering array! Black, brown, red, yellow and white—African, Amerindian, Asian, European, Eurasian, Polynesian—Ainu, Basque, Bushman, Chinese, Ethiopian, Lapp, Magyar, Malay, Masai, Pygmy, Turk—all peoples of earth yet disparate in culture, origins and beliefs. In what ways are they alike? In what ways do they differ?

Eric Severeid describes the task and the problem.

> "Only in the laboratories of science is the world truly one; in the laboratory of the mind and heart it is many worlds. People are differently spaced in the time channel of history, and they come to maturity and wisdom (or weariness and decline) at different moments, so that the problem rests in the fact that separate eras must live together within the one world."*

Perhaps it is in the laboratories, the libraries and classrooms on college and university campuses all over the world that the peoples of earth will best learn to understand each other.

---

*Eric Severeid, *Not So Wild A Dream,* New York: A. A. Knopf, 1946.

Chapter XII

ADULTS ARE ADULT

Teaching: Crux of Continuing Education

Ever since the mists of Pre-history lifted, the role of the Teacher, in most human societies, has been exalted and invested with an aura of dignity and importance. True, until modern society awarded him a specialized and prestigeful professional status, the Teacher very seldom was given the visible identity and ceremonial trappings of the Chief, the Warrior, or the Medicine Man (who in turn often combined the roles of Priest, Wise Man, and Physician). All of these assumed some of the functions and responsibilities of Teacher, as did the Parent.

An interesting parallel exists in contemporary society. The teacher of adults infrequently achieves general recognition as a fulltime professional. Many teachers of adults are primarily engineers, physicians, technicians, legislators, lawyers, farmers, bankers, housewives, clerks or members of other vocational groups, who assume the roles of a part-time teacher of adults in addition to their basic occupations.

One of the developments in American continuing education is the growing professionalism of administrators. One may infer that this means an increase in planned university courses of study leading to advanced degrees in continuing education. This may or may not be a desirable development! Observers of teacher education and of education of educational administrators might be forgiven for exclaiming, "God forbid that educators of adults be forced to lie on the Bed of Procrustes as are elementary and secondary school educators!"

It might also be said that the expansion of university preparation for continuing education is directed mostly at administrators. The role of teacher of adults is too often a part-time, occasional role. Many teachers of adults have little or no training for the difficult and complex task of teaching adults. Most have only a mastery of the tiny segment of knowledge which is to be taught by them in the usual standardized evening class meeting two hours one night each week for six or eight or ten weeks.

One of the needs of continuing education is greater provision for programs of in-service education for the teachers of adults who are employed by a myriad of diverse agencies and teach approximately 25 million adult students annually. Special training is often needed even for university faculty members who are recruited to teach adults as an occasional assignment in addition to their usual duties. Only a small fraction of public school and volunteer teachers of adults would probably be interested in degree oriented or college credit courses. Most of the in-service education programs should not even be on campus—they should be in the communities in which the non-college teachers live and teach. Specially designed programs of teacher education are needed—programs which encapsulate pertinent information about how adults learn, adult motivation, simple effective teaching methods, use of materials and teaching aids. The colleges and universities in the United States could make a significant contribution to improving the quality of adult educators if they would seriously devote resources to this task.

Teaching is both a science and an art. Teaching requires proficiency in skills, adeptness in a specialized craftsmanship, thorough knowledge and understanding of subject matter, a mastery of principles of learning, an innate rapport with people and a devotion to the task of teaching.

Few human activities are as demanding or as rewarding. It is significant that many of the great men and women of history have been teachers.

Much has been written about motivations of adult learners but little has been said about the importance of the motivations and attitudes of the teacher of adults. Barzun writes about the "dedicated teachers," the "fringe professionals" and the "campus followers." Most teaching of adults in the United States is done on a part-time basis—supplemental to the primary vocational activity of the teacher of adults.

Modern research has destroyed the myth of inability of adults to learn or be taught. But not all teachers of adults differentiate between intelligence, intellectual ability and wisdom.

An article "Adults are Adult," published in *Adult*

Leadership, spotlights the qualities of adulthood which are peculiarly vital in learning situations.

> "Adults are different. . . . Educators are aware, however, of the many ways, subtle, or obvious, in which adults while differing from each other, also differ markedly from children and from adolescents. It is recognition of these factors which so often spells the difference between failure and success in working with adults. . . .
>
> "Most adults have completed their formal institutionalized education. However varied might have been the extent and pattern of their formal schooling, most adults are engrossed in the tasks of living, and systematic visits to classrooms are adventures entered into reluctantly. Participation in adult learning activity may spring from deep wells of inadequacy or frustration—from a drive to succeed—from a great number and variety of motives and reasons.
>
> "Adults acquire vocational competencies. By the time the average adult reaches the middle years he has become skilled in at least one vocation. Many acquire supplemental competencies in hobbies and avocational pursuits. . . . Eventually his appetite for richer and more varied intellectual diets asserts itself and he welcomes the challenges and enjoyment for pursuing planned adult learning programs.
>
> "Adults accumulate experience. Springing from all of life's interests and events, and from the varied roles assumed on the shifting stages of life, such as marriage, home life, working for a living, sharing in community affairs; experience is, in reality, a kind of continuing education. Because of experience, adults need to be extended, by the teacher or director, all the rights and

privileges of full partnership in the learning experience and other adult enterprises.

"Wisdom is almost always a characteristic of maturity. It is the wisdom of the adult which enables him to contribute richly and uniquely to a group or community. Wisdom is not mere sagacity alone; it manifests itself in understanding of people, a perspective of life's values, sound judgment, insights into life's perplexities and sympathy for human weaknesses and shortcomings. The wise leader of adult groups attempts to mine this rich lode. Here again, age is superior to youth and the wisdom of a group of adults offers to the discerning teacher extraordinary challenges for exploitation. . . .

"Some adults find it difficult to maintain a scientific attitude. This is a negative factor, but a real one. To accept new ideas, to discard treasured concepts, to maintain a judicial and composed attitude toward change becomes more laborious as some adults grow older. . . .

"Here the teacher and leader of adults must call upon his acquired skills and his experience, probe the reservoir of his wisdom; for he too must measure up as an adult.

"Adult educators work with adults. They must therefore be sensitive always to the characteristics, drives, and yearnings of adults which determine the nature and quality of their participation in a learning group or in a community situation. Adult educators can never forget that—*Adults are Adult!*"*

It would be so easy, so simple to provide teachers

*Robert E. Sharer, "Adults are Adult," *Adult Leadership*, November, 1958, Vol. 7, No. 5, Washington: Adult Education Association.

of adults with special training for their roles and tasks, if there were a generally accepted, scientifically demonstrated theory of learning of adults. There is no single theory of adult learning.

Just as educators disagree about theories of adult learning, so they differ about the effectiveness of techniques and methods. There has been a proliferation of techniques, as research and the technology of communication move forward. There is need for wider use of some of the new patterns and forms which experimenters have described. *Electronics may yet challenge tradition, and the much used lecture method may lose its popularity to some new technique which combines personal presentation with effective learner participation. A whole new challenge to older methods is unfolding as programmed learning, and automated-instruction spreads into adult education. The "modern" teacher of adults needs to be an avid reader of research in adult education and an experimenter with new techniques to maintain his preeminence.*

Instruction—the art of teaching—remains the crux of adult education. All factors such as preparation, experience, personality, subject matter, techniques and teaching skill reinforce the centrality of instruction.

No other individual in the entire galaxy of educators can equal the teacher in the importance of function or in the vital significance of contribution to either individuals or to society.

Chapter XIII

CREATIVE PROGRAM DESIGN

The Changing Spectrum of Continuing Education

No man achieves his fullest potentiality without a challenge. One of the objectives of continuing education is to supply the individual with a challenge, or a series of challenges, to grow intellectually and spiritually until he achieves his ultimate potential development.

Since no two individuals are exactly alike, or respond alike to any challenge, or set of challenges, it is a tragic error to routinize and standardize programs of adult education. Nor is it sufficient to invent bizarre or exotic titles to disguise dull and uninspiring programs.

Since society is in a constant state of change and upheaval, it is equally tragic indefinitely to continue unchanged programs of continuing education which supposedly are geared to societal needs and problems. Our mass society tends to force all individuals into identical molds. Continuing education can, should and must encourage and stimulate and preserve individualism. It is significant that some object to the term "continuing education" on the basis that it connotes more of the same standardized educational formats and processes employed in pre-adult education, which destroy creativity, individualism and independent thinking.

An adventure in continuing education begins with designing programs. The change from an idea to a viable program is like a metamorphosis. In the designing and operating of any continuing education program the educator can use any residue remaining in his blood of the explorer-pioneer-pathfinder spirit of his forbears. This is the opportunity to couple imagination, ingenuity and inspiration with common sense, sound knowledge and skill.

What is an education program? Some regard it as essentially a transaction between a client and an adult educator. Even if the criticism that it connotes too much emphasis upon the economic procedures of the program is disregarded, this definition is incomplete.

Another traditional definition, somewhat trite and perhaps very pedagogical is: "teachers teach subjects to students." Although this statement emphasizes instruction, it ignores learning situations where there are no teachers, or instances when subject matter is relatively unimportant.

Designing and launching a program of continuing education can be compared to a complex reaction, involving a number of variable components, each interacting with the others, and producing a program which results from the combination and interaction. An equation representing this can be written thus: "clientele plus purpose plus content plus instructor plus techniques, plus format plus setting plus sponsor plus procedural factors plus catalyzer (creativity of educator) produces—a program of continuing education."

Of course the adult educator does not measure out certain amounts of each ingredient and mix them thoroughly in a specially designed test tube. But he needs to understand all these factors, and their potential contributions to the programs which he plans and hopes will be a success.

Creativity in program design can be achieved in some degree by deliberate imaginative alterations in the form or character of any one or several of the ten components in this process. It should be pointed out that there is a certain degree of unpredictability in the interractions among the several components, and that a certain amount of the spirit of adventure should possess the soul of the program entrepreneur as he launches each program.

Let no one be misled into the blind alley of thinking that creative designing of programs is a mechanical procedure, or the following of directions of a recipe, or a laboratory experiment with reagents and reactors. Essential vital factors are the imagination and professional skill of the program designer.

There are many varieties of groups for which programs are designed: farmers, bankers, corporation executives, government officials and organization leaders. A program might be planned for parents, or voters, or teachers, or housewives, or ministers. Occasionally the clientele of a program represents many types and kinds of person, brought together because of a common interest. It is almost trite to point out that the needs and interests of the adult learner are and must always be paramount in the designing of programs of education.

Continuing education programs have purposes and goals. One is universal—to satisfy the specific needs and interests of adults. Sometimes the goals are progressive—the program points the way and leads the learners to desire expanding and advancing goals, leading into additional programs. There are usually purposes of the sponsor which need consideration. Programs usually are appropriate to the basic and more general purposes and concerns of adult education. When they are not compatible with the basic goals of adult education, the results are usually unsatisfactory.

The substance or content of a program is usually its subject matter, or the skills to be acquired. Sometimes programs are listed in terms of traditional subject matter or disciplines, such as history, mathematics, chemistry, anthropology and typewriting. It can be based upon broad social needs: literacy, health, aging, family living, public responsibility or world affairs. The content may be devoted to acquisition of skills: public speaking, discussion leadership, group processes. A program might be concerned with fact-finding through research or surveys.

Certainly the choice of instructor or program leader is a key decision in program planning. The competency, enthusiasm, personality and attitude of the teacher can determine the success or failure of almost any program. Successful teaching of children or adolescents does not always guarantee successful teaching of adults.

There are many program forms which represent the popular formats of continuing education programs. With over a score of proven formats to choose and no door to experimentation closed, the designer should never fear monotony in this aspect of program design.

Techniques and methods are sometimes confused with program formats. The two most used techniques are the traditional classroom lectures and group discussion. Both, although widely used, vary greatly in effectiveness because of differences in abilities and competencies of the lecturers or discussion leaders.

The settings of continuing education programs vary widely from a living room in front of a television set to a well equipped university auditorium. It might be a quiet retreat on the ocean shore or in a mountain valley. It can be a public school classroom, a library reading room, a labor union hall, a church Sunday school room or a hotel banquet room. The spread of residential conference facilities on college and university campuses during the past ten years has been remarkable. Architects of school buildings, university buildings, libraries; even factories and hospitals have begun to incorporate facilities for adult learning groups in their designs.

The sponsor of continuing education programs may influence the program or there may be definite organizational or institutional policies which limit or modify its programs. Continuing education programs sponsored by an organized labor group would not be expected to present points of view antagonistic to labor unions. National and state farm organizations have their official policies in regard to problems and issues.

Procedural factors, including costs, registration procedures, prerequisites and length and number of meetings are not inconsequential. Too often the inadequacy of physical facilities adversely affects the adult education enterprise. Such simple items as distance between home and the program and transportation facilities need to be considered. Some decry the influence of "Madison Avenue" on adult education, but the quality and effective distribution of information about adult education offerings is important.

Aleatory Programming is not unexpected and a most intriguing development of Cybernetics. The composing of both solo and orchestral music scores by means of a computer has produced arrangements which have attracted public interest. Business and industry are experimenting with the use of computers in design of products and sales campaigns. Educators are being told that ten or a dozen carefully selected factors which control the character and format of a continuing education program can be fed into a computer with the purpose of creating a unique and usable program. It is not the substitution of mechanical-electrical devices for the alembic of human creativity which is so upsetting. This idea raises the specter of a continuing educator in the future. A robot electronically directing programs of continuing education based upon instructions supplied by a computer!

Who can estimate the effect of the enthusiasm and ingenuity of the adult educator on the program? How about the personality, the inventiveness, the contagious love for knowledge of a teacher of adults?

The task we have been describing in this chapter, the task of creative design of adult education if successfully discharged, results in unleashing uniquely human creative contributions to society. This is a basic goal of continuing education.

Chapter XIV

BRICKS AND MORTAR

Procedural Matters in Program Building

Administrators of continuing education have been characterized as artists, or as philosophers, perhaps even dilettantes. Resemblances have been noticed because the administrator deals with ideas and concepts, and with intangible services, fabricated from the stuff of his dreams and aspirations. But the administrator with equal appropriateness can be described as an artisan, a builder, an engineer or an architect. The bricks and mortar and building stones with which he fashions educational programs with skilled, creative craftsmanship are composed of items such as finance, marketing, staff, inter-agency relationships, institutional policies and institutional involvement—all the functions and concerns of administration and management. These are the procedural affairs which even the most adventuresome educator cannot ignore or forget.

A complete discussion of the complex and varied procedural affairs in continuing education would include examination of research in continuing education, evaluation, procedures uniquely essential for conferences, workshops and residential learning situations, adult educational publications, planning and use of educational facilities of physical plant, special library facilities for adult learners and correspondence and home-study programs.

Let us confine our discussion to four procedural concerns for consideration: financial procedures, advertising and marketing procedures, interagency cooperation and collaboration and procedures of internal organizational operation.

The financing of continuing education is complex and confusing. There is no uniformity among agencies on principles or patterns. Currently four major resources are employed to finance adult education in the United States. Only a few programs utilize only one of these; most programs are financed by a combination of at least two, sometimes of three or even of all four methods. They are program cost assumed by the clients, program cost assumed by the sponsor, program cost assumed by local, state, or federal government or by combined support from two or three of these sources and program costs assumed by grants from foundations, corporations or individuals.

Privately owned mass media agencies—newspapers, magazines, radio stations, television stations—have steadily assumed larger roles in programs of public information and education, without direct charges to readers or viewers. As trustees of important adult educational responsibilities, these agencies may assume even larger roles, and questions of financial support will be raised.

Marketing or merchandizing (some educators wince and shudder at the use of these terms in application to education) is another important procedural concern. The influence of "Madison Avenue" is being felt by American educators; some are yielding, some are resisting its influence. Those who are yielding are trying to re-design their public information, promotion, publicity and advertising techniques. Others maintain that programs based upon real individual and social needs do not require high pressure advertising techniques. Some argue that if adults are involved in planning and evaluating programs, extensive advertising is unnecessary. Still others contend that Americans are conditioned to "Madison Avenue" techniques and, unless continuing education employs similar methods, will be unimpressed and fail to attend. There are many sincere educators who fear "hucksterism" in education and are extremely reluctant to accept some of the suggestions of promoters and advertisers.

The educator as he seeks procedures and methods of acquainting adults with his programs and services seeks ways which are effective, truthful, dignified and in good taste.

Interagency reciprocity, an active sharing, interdependence, exchange, and cooperation among the many agencies, institutions and organizations which engage in continuing education, has been a dream of American adult educators. Problems of effective cooperation are multiplied or made more complex because the several agencies are not only diversified in kind of type, but operate on various levels—the community, the state and national level. Rapid development of effective procedures and devices to increase cooperation among adult education sponsors has been hindered in many states and communities by apathy, jealousy, fear of loss of status, interagency rivalry and confusion over roles and responsibilities.

In the United States there have been many successful ventures in cooperation and coordination among continuing education program sponsors. Six types of varieties of planned cooperation can be discerned; informal interagency cooperation in neighboring communities, adult education councils in communities, intercommunity agency cooperation, state institutional cooperation, state and local agency cooperation and national level cooperation.

A fourth major complex or set of procedural affairs may be described as intra-institutional or intra-mural affairs. Within the institutions and agencies which continuing education has stimulated, a number of bothersome problems and issues have developed. The procedures and "methods of getting things done" have become complex and many times frustrating in the university, the public school and in institutions which operate continuing education services.

Continuing education in many agencies continues to be a marginal activity, despite its growth and the general acceptance by educators of its social importance. This relegates to a low priority all institutional allocations of budget, personnel, facilities and research. The person chiefly in charge of the continuing education arm of the institution finds it necessary to devote considerable time and effort to battle for recognition within the institution and to secure the allocations of money and faculty he needs.

A second source of administrative problems is the body of total institutional traditions and policies which affect every institutional program. How can the program director of a state Farm Bureau plan a completely unbiased series of study—discussions on national agricultural problems, when his organization has definite official points of view about these issues? How can a public school director of adult education plan for community-wide study of needed tax structure reform when his board of education opposes any reduction in school tax allocations because it will jeopardize the school building program?

In most institutions there exists what John Dyer calls "power on the campus." It may not be an institution of higher learning, but it will have a typical "power structure" in its organization. The educational administrator must be alert to, and skillful in dealing with those within his organization who direct and control. One of the often forgotten roles of an educator is that labeled by Leland Bradford as a "keeper of the keys." If he holds the keys to internal relationships within his organization, the educator can unlock the doors and open the gates to program innovation and expansion. Otherwise he may remain within his institution, "cribbed, cabined, and confined," destined to be a kind of glorified clerical employee.

Still another aspect of internal procedural affairs is concerned with the recognition and utilization of volunteers and "non-professionals" by adult education agencies. The trends toward greater professionalization of staffs, the institutionalization of programming and emphasis upon adult education as a bona fide business enterprise have created a number of consequences. The occasional "do-good" volunteer excursions into continuing education seem to be handicapped by the more efficient, more polished activities of the big institutions and organizations. This is to be deplored because continuing education has a largely unexplored and unused resource of tremendous potential among the many thousands of adults who could be utilized as instructors, leaders and evaluators in adult education. Cooperative Extension has made good use of the volunteer educator, benefiting both the agency and our rural population.

A few administrators of continuing education are the chief executives of a total enterprise devoted entirely to adult education. The executive officer in this ideal arrangement finds himself in a dual partnership. He is a partner in the policy making and managerial affairs with his governing board or committee. He is also a partner with his staff of colleagues who are responsible for implementation and operational procedures.

Problems of personnel relations and personnel management are discussed in professional literature. Perhaps the continuing educator in staff and faculty relationships will discover many interesting parallels to those which exist in business, in industrial establishments and in universities and colleges. Most of his concerns will be in selection of staff and faculty, in-service education of staff and faculty and organization of staff and faculty.

His major concerns will be threefold: How can maximum effectiveness be secured? How can creativity be encouraged, yet controlled? How can leadership in adult education be maintained in spite of strain of administrative or instructional duties?

The competent skillful administrator may, like the captain of a ship or the pilot of a plane, accomplish some degree of success in these personnel procedures by instinct and innate wisdom. Much will have to be acquired from experience and study. The weltering complex of procedural affairs constitute a phalanx of antagonists which must be faced by the adult educator as he, each day, enters the tournament of his job.

Procedural matters—routinized and formalized, or experimental and innovative—can become a morass which entraps the educator and compels him to expend his energy in attempts to extricate himself and his program.

Or, as indicated in the opening paragraphs of this chapter, procedural affairs can be building materials—the bricks and mortar and building stones out of which the adult educator constructs the edifice of his ideal program.

The administrator must have many competencies. He is a composite of idealist and activist, of poet, artist, philosopher, of architect, artisan, engineer and builder. He uses practical procedures and materials to build stately mansions for the mind and spirit of his fellow men.

The administrator of continuing education is a creative builder in society. I know of no other occupation which has greater opportunities for accomplishments of timely and timeless worth.

Chapter XV

A COMMUNITY OF DIVERSITY

Can Continuing Education Achieve Its Destiny?

Education, including continuing education, is an important invention of man. All of its glories and all of its weaknesses may be ascribed to its inventor.

"When we look realistically at today's world and become aware of what the actual problems of learning are, our conception of education changes radically. Although the educational system remains basically unchanged, we are no longer dealing primarily with the vertical transmission of the tried and true by the old, mature, and experienced teacher to the young, immature, and inexperienced pupil in the classroom.

"What is needed and what we are already moving toward is the inclusion of another whole dimension of learning: the *lateral* transmission, to every sentient member of society, of what has just been discovered, invented, created, manufactured, or marketed.

"Thus we avoid facing the most vivid truth of the new age: *no one will live all his life in the world into which he was born, and no one will die in the world in which he worked in his maturity.*"*

---

*Margaret Mead, "A Reflection of Education," *National Education Association Journal*, Washington, October, 1959.

If man is "the glory, jest and riddle of the world," then his inventions may well reflect his qualities and traits.

There is little doubt that an extra-terrestial visitor would regard the diversity of sponsorship and programs in continuing education as its most remarkable characteristic. He might logically ask: is this an advantage, or a handicap; a potential source of strength, or an inherent weakness? American educators, as they study its origins and history, inventory and appraise its goals and accomplishments and view its contemporary strengths and weaknesses against the backdrop of our contemporary society, are prone to ask the same question.

Because of its great diversity, a hurried look at continuing education is certain to produce discouragement, if not dismay. From whatever viewpoint we examine continuing education; in terms of sponsors, or content, or methodology, or functions, or purposes, we find variety and diversity.

But diversity is not peculiar to continuing education. Our historic American accent on individualism, free enterprise, ingenuity and inventiveness has resulted in diversified economic, social, business, political, governmental and religious structures in our society.

We are prone to forget that American adaptiveness has made this characteristic of diversity a source of strength and great achievement. By cooperative endeavor and teamwork, individuals, groups and divergent interests have pooled their resources, united for common purposes and made valuable contributions to community and national welfare.

There are many evidences in the loose, fluid structure of continuing education of the application of this American tradition of cooperation. There are many indications of far more than fellowship among adult educators; there are indications of a "community-ship."

The massive proportions of continuing education are a challenge to united effort. We are in the midst of a continuing education explosion in the United States. No other descriptive term is adequate. Estimates of numbers of adults participating last year in planned programs range between 25 and 75 million. And the tidal wave continues to mount.

The essential character of adult education is becoming more acutely geared to the circumstances and issues of our times. It is easy to find evidences of this development.

For 50 thousand years men walked from place to place, or rode in ox carts, or on backs of camels and horses. It has been only 100 years since steam revolutionized travel on sea and land. Then came the automobile and, in this twentieth century, the airplane.

Jet planes traveling faster than sound went into the skies as World War II ended. Now planes cross oceans and continents at 2000 miles an hour and our space ships encircle the globe in 90 minutes. Conservative scientists predict that a resident of the earth will land on the moon, perhaps within two or three years.

It is entirely possible that within the lifetimes of many of the inhabitants of this earth man will learn if intelligent beings exist on other planets in this universe.

The impact of all this has created new concepts of human life and human purpose. These in turn are transforming traditional concepts of continuing education.

Educators are adapting technological advances to the methodology of continuing education. Development of radio and television as instruments of education has been phenomenal. A favorable attitude has developed toward experimentation with automated learning, use of teaching machines, the application of findings of research in communication and the adaptation of techniques perfected in the area of business and industry.

Research in adult education is no longer regarded as of secondary importance as grants from major foundations indicate. Before 1885, there were no sociologists, no psychologists on our college faculties or listed in "Who's Who!" Colleges of Engineering and Colleges of Business have achieved university acceptance since 1900. "Adult education" as a term did not appear in U. S. educational literature until after World War I. Until recently most individuals engaged in continuing education have entered the profession after preparing for some other occupation.

We have a new generation of able young men and women, educating themselves in carefully designed undergraduate and graduate courses of study for years of professional activity and productivity in continuing education.

Co-planning and coordinating procedures and devices have multiplied. This development is evident in communities, on the state scene and on the national level.

The proliferation of state associations for continuing education is evidence of common desire among adult educators to pool their ideas and efforts. Although these organizations have many goals and purposes, their most significant function is providing opportunities for cooperation.

Chapter XVI

A STATE ADULT EDUCATION ASSOCIATION

A Suggested Paradigm

There are many forms and kinds of state adult education organizations. This situation reflects the diversity and complexity of U. S. adult education. Not all state associations are effective. A number of factors contribute to this. Adult educators, who are planning a new organization, or are engaged in a revision of their existing organization, should carefully examine the several factors which enhance or inhibit the effectiveness of an organization.

The following outline is not intended to be a blueprint or a model. It is an outline of suggested items which might be thought of as a paradigm or example in designing a state association. Each item is subject to review, modification, rejection, or acceptance, and additional items very appropriately might be added.

I. Purposes

1. To provide a practical, workable framework which will enable individuals and groups to join in a common purpose, when the need arises.
2. To provide effective leadership, supported by

informed and concerned followership, and to encourage desirable courses of action.

3. To provide opportunities for adult educators and those interested in adult education, although representing widely diverse interests and programs, to share and exchange ideas; to explore common problems and desirable solutions; to become personally better acquainted with each other; and to substitute a realization of unity or purpose for individual anomie.

4. To focus the attention and concern of educators and the public upon the major issues and dilemmas of our time, and suggest ways and means in which adult education can contribute to finding the most satisfactory solutions.

II. Members

1. Individuals whose major occupational responsibilities are in some form of adult education.

2. Individuals whose occupation is education, but whose responsibilities in adult education are minor, or who wish to give support to adult education.

3. Individuals whose occupation is other than education, but who believe in the importance of adult education and wish to give effective support.

4. Individuals engaged in any occupation or profession who serve on a part-time basis, as volunteers in adult education.

5. Institutions, agencies, government departments and/or organizations which desire to support adult education.

6. Affiliates: organizations wishing to cooperate but whose charters prohibit membership in other organizations.

III. Activities

To implement the purposes of the organization, the following activities should be planned and carried forward.

1. To provide sustained, carefully planned public visibility for the major educational and civic needs in the state.
2. Seek financial support from private and public sources for programs and projects which are urgent, and which the organization can best undertake.
3. Seek acceptance by private and public institutions and agencies of responsibility for initiating and conducting important studies, programs and activities which are appropriate for them.
4. To develop newsletters and other forms of publications in which members might present points of view, or describe successful programs.
5. Cooperate with the Adult Education Association, USA.
6. Conduct area and state conferences.

IV. Organizational Machinery

1. Devise, and periodically modify a simple, administratively flexible charter or constitution.
2. Democratically elect able officers, entrusted with considerable latitude, and given opportunities to exercise intelligent leadership.

3. Provide ample opportunity during scheduled conferences for thorough discussion by the membership of organizational policies and activities.

4. Provide for a balance of powers in decision making among the diversified element and occupational groupings represented in the membership.

5. Insure periodic rotation of leadership responsibilities among those qualified.

Note: This statement was prepared in 1965 for the organizing committee of the Adult Education Association of Wisconsin.

Approximately 15 national organizations in the United States appropriately might be labeled as supportive of continuing education because of their major purposes. Some, like the National University Extension Association, the Association of University Evening Colleges and the American Association of Land Grant Colleges and Universities are based upon institutional membership. The National Association of Public School Adult Educators is based upon individual members, but restricted to public school personnel. Still others, such as the American Library Association, embrace adult education as one of the several major concerns.

Among national adult education associations, the Adult Education Association—U.S.A. is unique. It is the one primarily individual-membership organization. Some regard the AEA-USA as a kind of service station or supermarket with free services of all kinds! Others think it should be modeled after the monastery—a restricted community of specially selected scholars and devotees of adult education! I have heard the association likened to a dynamo—a device for renewing enthusiasm and professional vigor—an educational battery charger, as it were!

The AEA-USA has several important uses. It serves as the "uniter" for all organizations when emergencies call for united study or concerted action. It serves as a national clearing house and information exchange. It lends support to state and local adult education associations. It can assume the role of catalyst in the capital city of our nation stimulating interest and action in political and educational agencies.

Its ultimate fate—the fate of all 15 "national organizations for continuing education"—will be decided by the relentless forces of time. Perhaps someday the U.S. will have a Federation of Adult Education Associations, if not a single unified organization.

The main question is *what is the destiny of continuing education?*

The ultimate character and significance of continuing education cannot be predicted with any degree of certainty. It is certain that many, many individuals of diverse backgrounds, interests, vocations, needs and concepts will participate in the determination of the ultimate character of American continuing education. Shaping and molding continuing education will be all the complex societal forces which continually dynamize human societies.

There will always be a great need for able leaders—men and women of intelligence, wisdom and vision. There will always be a need for educational trail-blazers and pioneers who will attempt to chart new directions for adult education. There will always be a need for devoted, competent teachers of men and women, skilled in use of all the new technological devices and the application of the findings of research. And, of course, there will always be a need for able administrators who possess the executive ability, who really believe in the great importance of continuing education, and who have the courage of adventurers!

Chapter XVII

TAR 'N CEMENT

An Unrealized Dimension of Continuing Education

It was a beautiful Spring morning on the campus. In the Student Union Grill a handful of students and faculty dawdled over late breakfast or mid-morning coffee. Then a student arose from his table, walked across the room and dropped a coin into that automated dispenser of syncopated sound, popularly known as a juke-box. The no longer quiet room was filled with sound.

"Where are the meadows?
    Tar 'n cement
The laughter of children?
    Tar 'n cement
Where is the tall grass?
    Tar 'n cement
Nothin' but acres of tar 'n cement."*

Long after I had left that room the lament continued to haunt the recesses of my mind. Through the open window, all day, came the continuing swish

---

*TAR AND CEMENT. (IL RAGAZZO DELLA VIA GLUCK). English words by Paul Vance and Lee Pockriss; music by A. Celentano. New matter: English words. Registered in the names of Edizioni Musicali Clan and Edizioni Curci, under E unpub. 929092, March 15, 1966. The Library of Congress. Recorded on Imperial Records, Mel Carter, soloist.

of tires on tar and cement, punctuated by the occasional squawk of an auto horn as a weary driver vented his irritation.

Is this the doom of twentieth century man? All day the phrase "tar 'n cement" seemed to describe all too clearly the character of the existence man endures.

Later, in the quiet evening hour, I opened my notes and began to read the introduction originally planned for this chapter. I read: Centuries ago Aristotle wrote, "There is a life which is higher than the measure of humanity. Men live it not by virtue of their humanity, but by virtue of something in them that is divine."

A twentieth century scientist writes in an American periodical:

> "Life defied measurement. Only the properties of nature, not the essence, can be described in quantitative terms. We can weigh a mouse and determine the length of its ears and tail; we cannot measure the quality of a mousehood. With elaborate scientific devices we can explore the physiology of the human organism, but our devices are yet powerless to reveal the essence of humanness."*

Like a flash the thought came—it's all there—the answer is right there. It all depends upon how the individual regards his most precious endowment. Is it a mere existence or is it a life to be lived?

And right there was the answer to the related question—"For what purpose education?" If education is designed to prepare man to endure existence

---

*E. L. Grant Watson, "The Hidden Heart of Nature: Adventures of the Mind Series," *The Saturday Evening Post*, May 27, 1961, Philadelphia: Curtis Publishing Company.

in a world of noise and smog and machines, then the whole character of education is determined by what purpose.

But, if education is to assist man to achieve the fullest potentialities of life as perceived by Aristotle and Watson and other great thinkers, education then assumes a very different quality.

Although separated by a gulf of 25 centuries of time, the Greek philosopher and the American scientist each, in a flash of insight into the nature of human life, speaks of the miracle and potentialities of the human mind and of the sublimity of the indomitable spirit of man. In this area of mind and spirit lies a tremendous challenge and opportunity for adult education in the United States. New meaning and added significance for continuing education are encompassed in the phrase education for qualitative living.

Historically continuing education has been mostly occupied with the quantitative aspects of education. The acquisition of facts, the learning of skills, the invention of machines and gadgets, the requirements for earning a living; these have constituted the major concerns of adult programs of education, as they have comprised most of the content of all levels of education. This preoccupation with the "business of living," this slighting of the "art of living" has created a sector of our population which is strangely illiterate in certain qualitative aspects of education, while possessing great proficiency in the quantitative areas of knowledge and skills. It is not surprising that a concern with greater emphasis upon the orientation of man as man rather than man as a money maker has been receiving more and more attention during this decade. Men and women are beginning to believe the statement of H. G. Wells: "Knowledge is the only way out of the cages of life."

Education for qualitative living may be projected in four dimensions.

Education for individual improvement and growth includes programs which emphasize the acquisition of new insights and points of view. This dimension includes: the examination and consideration of values, including ethics, morality, ideals and religion; and the study of human behavior with special attention on character, standards of conduct and attitudes. Serving as reservoirs of ideas for this entire area are the disciplines of humanities, history, religion, philosophy, psychology and the behavioral sciences.

The second dimension of education for qualitative living is concerned with greater understanding, appreciation and enjoyment of the arts and of the sciences. This dimension is usually intended by those who use the term Liberal Arts adult education. Dr. William Birenbaum, Vice President of Long Island University, presented a paper before the faculty of the Graduate School of the U. S. Department of Agriculture on March 14, 1963. His concluding paragraphs prophetically picture a growing realization of the importance of this aspect of continuing education.

"In an automated, highly organized, urban epoch, the human capacity to know beauty may turn out to be the quality most useful to the maintenance of sanity.

On a crowded and explosive planet the aesthetic wells of the human mind may be the most precious reservoirs of the human spirit. In a day-to-day life of traffic jams, assembly lines, shortened work-hours, and enlarged purchasing power, the cultivation and nourishment of the artistic impulse in every man may be the most practical and promising of all human pursuits.

"Indeed in the context of modern life, the futility of utility, as we now see it becomes increasingly apparent. The great revolutions of the Twentieth Century do not have to do with the rise and fall of governments, but with the fall of man's traditional view of his limitations, and the rise of a renewed elaboration of his possibilities. Each new discovery cries out for poetic description. We may be the lucky ones. We may be the poets of the human adventure."*

The third area of education for qualitative living embraces creative experiences, the actual kindling of hidden talents into imaginative original contributions to our culture and to the lives of those who create, make, design or compose. Adults are encouraged to write prose or poetry, to compose music, to engage in painting, drawing and sculptoring, to devise architectural plans, to write plays. American adult educators have been slow to realize the implications of this dimension of continuing education, not only for the development of American artists, writers, composers, architects, poets and dramatists, but also for its potential contributions to American culture and life.

---

*_The Graduate School Newsletter_, April 15, 1963, Washington, D. C.: U. S. Department of Agriculture.

No discussion of education for qualitative living is complete without a reference to still another dimension of this essential form of education. This fourth dimension of education (no reference to space mathematics is intended here!) is based upon a concept of education very different from those which are the foundation of the three dimensions described earlier in this chapter.

This dimension is sometimes called "liberating education," because it is primarily concerned with a process, not with mastery of skills, or areas of knowledge. The whole concern of education is transferred from what is studied, to what is happening to the learner. The perspectives and understanding which are the goals of "liberating education" aim at breaking the bonds of prejudice, bias, ignorance and fallacious concepts which so often encumber the hearts and minds of adults. *Education, concerned mostly with the quantitative aspects of living is tremendously important in our technological age, but is is entirely inadequate and impotent in the task of helping man free himself from he restraints of a monolithic society; from intellectual myopia which is the occupational disease of over-specialization; from the narrowness of perspective and understanding of meanings and events of life.*

There are educators who regard these dimensions as the most important of all developments in continuing education. Others, perhaps more timid or tradition-bound, fear that emphasis upon this aspect of adult education tends to encourage rebels, heretics and intellectual malcontents. There are among educators many pragmatists who label education for qualitative living as poetic, impractical, academic and not wanted by the average adult. There also are educational leaders who deplore these fears and reservations, and point to the importance in our society of pioneers and trail blazers in human thought, of innovators in educational practice, of experimenters in all the fields of human knowledge, of explorers in concepts and principles of human existence.

It would be a tragic development in this critical episode of American continuing education if its leaders, administrators, researchers and instructors dedicated most or all of their strength and effort to education for the quantitative aspects of human life. It would be tragic if they neglected to devote some fraction of their creative leadership to expanding and making more effective the very difficult but important aspects and forms of continuing education which contribute to qualitative living, including the four dimensions described in this chapter. Our institutions of education must invest some "risk capital" of administrative leadership, research and financial support to the development of education for qualitative living.

Some of the most discerning educators have joined with historians, with observers of the kaleidoscopic events occuring on the national scene, and with public leaders in asking "Will America share the destiny of Rome?" This should not be considered merely a rhetorical question. The destination toward which some think our nation appears to be traveling inexorably tugs all levels of education in the same direction.

This situation might become sufficiently disturbing to embolden educators to accept a new concept of continuing education, a new basis upon which could be built a "new continuing education" in the United States. This *new* continuing education would include programs of occupational education concerned *not only* with avocational woodworking, cake decorating, eradicating potato bugs or operating a drill press, but also effectively aid adults in maintaining occupational competence in spite of accelerating occupational obsolescence. This new continuing education would emphasize as vital, integral components programs designed to assist men and women to achieve spiritual and intellectual competence, in order that they might achieve their destiny to be only a "little lower than the angels."

This concept is far more than merely education for leisure. Thoreau wrote: "We cannot kill time without injuring eternity." This is education to be continued during the hours not required to make a living. This kind of education may prove to be more satisfying and more productive than education only for the quantitative aspects of living.

Continuing education can never achieve its most significant contribution until it accepts the imperative of utilizing education to improve the quality of human life.

This is a larger task, a greater challenge than we

have ever attempted—to plan and design education programs for improving the quality of living. Are we intelligent enough? Are we imaginative enough? Are we inventive enough?

If we are somewhat doubtful, or timid, let's put a coin in the juke-box again, and listen:

"Where are the meadows?
Tar 'n cement
The laughter of children?
Tar 'n cement
Where is the tall grass?
Tar 'n cement
Nothin' but acres of tar 'n cement."

AFTERWORD

No reader is surprised to find a "Foreword" in a book but an "Afterword" is somewhat unusual. It provides the author with a final opportunity to summarize his points of view and to reaffirm one or two of his convictions.

The subtitle of *There Are No Islands* is "The Concerns and Potentials of Adult Education."

In the sixteen chapters positioned between the Foreword and this Afterword important *concerns* of American continuing education have been discussed. These may be summarized briefly as follows.

Determine desirable emphases, goals and directions.

Define, clarify and implement appropriate roles by agencies and institutions.

Examine and refine the substance of adult education.

Apply research and experimentation to the improvement of methods and techniques.

Develop effective preparation and in-service education of professional educators in continuing education.

Critically evaluate current programs and techniques on basis of socially and scientifically defensible criteria.

Adopt sound administrative and managerial principles and practices in continuing education enterprises.

Invent more effective community and area devices for coordinating continuing education agencies and programs in communities.

Increase amount and velocity of interchange of creative ideas and practices among adult educators in the U. S.

Promote international exchange of ideas, practices and philosophies of continuing education.

Safeguard and encourage the commitment and efforts of dedicated volunteer leaders and workers.

Discover, enlist and "set on fire" dedicated volunteer leaders and workers.

The *potentials* of American continuing education were given extensive consideration in preceding chapters. That much of consequence and significance remains to be attempted is clearly evident. The "yet to be accomplished" provides a greater incentive than the laurel wreaths of yesteryears. All educative enterprises including continuing education are future-oriented.

In a real sense all individuals are forever strangers in a forever new world, a world being renewed every day and vastly different every year. Leaders of continuing education must be alert, adaptive, insightful and sensitive to human needs and society's challenges. The contemporary revolution which American society is undergoing offers to continuing education a chance to participate in making history, a chance to help black Americans to prepare for the new obligations and new opportunities which they are demanding.

Another potential is the elimination of the "young adult gap" in the spectrum of continuing education. The involvement of young adults is program design and program leadership and participation of young adults in continuing education programs is far too meager. Continuing education and youthful adults will benefit when this has been achieved.

The utilization of the almost untapped reservoir of talent, wisdom and experience of retired leaders in community affairs, the professions and the business-industrial sector of society is a third major potential of continuing education.

To permit continuing education in America to become standardized, routinized and mechanical will destroy its unique appeal to individuals and emasculate its ability to benefit society.

Educators must never permit themselves to develop callouses on their souls, or to lose the dreams of their youth.